IDIOT'S GUIDES®

AS EASY AS IT GETS!

Calculus II

by Chris Monahan

ALPHA

A member of Penguin Random House LLC

Publisher: Mike Sanders
Associate Publisher: Billy Fields
Acquisitions Editor: Jan Lynn
Development Editor: Rick Kughen
Cover Designer: Lindsay Dobbs
Book Designer: William Thomas
Compositor: Ayanna Lacey
Proofreader: Laura Caddell
Indexer: Tonya Heard

This book is dedicated to one very special woman, my wife, Diane. Her encouragement throughout this process was always appreciated.

First American Edition, 2016
Published in the United States by DK Publishing
6081 E. 82nd Street, Indianapolis, Indiana 46250

Copyright © 2016 Dorling Kindersley Limited
A Penguin Random House Company
16 17 18 19   10 9 8 7 6 5 4 3 2 1
001-295791-DECEMBER2016

ISBN: 9781465454409
Library of Congress Catalog Card Number: 2016941029

Note: This publication contains the opinions and ideas of its author(s). It is intended to provide helpful and informative material on the subject matter covered. It is sold with the understanding that the author(s) and publisher are not engaged in rendering professional services in the book. If the reader requires personal assistance or advice, a competent professional should be consulted. The author(s) and publisher specifically disclaim any responsibility for any liability, loss, or risk, personal or otherwise, which is incurred as a consequence, directly or indirectly, of the use and application of any of the contents of this book.

Trademarks: All terms mentioned in this book that are known to be or are suspected of being trademarks or service marks have been appropriately capitalized. Alpha Books, DK, and Penguin Random House LLC cannot attest to the accuracy of this information. Use of a term in this book should not be regarded as affecting the validity of any trademark or service mark.

DK books are available at special discounts when purchased in bulk for sales promotions, premiums, fund-raising, or educational use. For details, contact: DK Publishing Special Markets, 345 Hudson Street, New York, New York 10014 or SpecialSales@dk.com.

Printed and bound in the United States of America

Contents

Appendixes

Introduction

If you look up the word *calculus* in the dictionary, one of the first definitions given is that it is a hard object, like a kidney stone. People used to joke that it was no wonder that the subject was so difficult to pass. This book is intended to "soften the calculus" so you are able to better understand Calculus II.

I have tried to avoid using the technical language of mathematics whenever I could. When I absolutely had to use more technical language, I have translated those terms into everyday language you can comprehend. Also, you'll find plenty of examples in each chapter to help you understand the solutions to each problem. These examples include easy-to-understand explanations as well as the requisite mathematical notations.

However, I warn you not to be a "mathematical voyeur," one who likes to watch the math being done but who does not do it himself. As you are working through this book, be sure to have paper, pencil, and your graphing calculator by your side—and use them. Read the example, do the problem yourself, then look at my solution. You might want to hide the example solutions before you do the problem so you cannot cheat. People who have taken calculus will tell you that the first step in the problem is calculus, but the rest of the steps needed to solve the problem are algebra. Be careful as you work.

Not every example problem included is easy. If I took that approach to teaching Calculus II, I would be giving you a false sense of what to expect. Making things too easy also would deny you the satisfaction of taking on a challenge and succeeding.

Finally, I used to tell my students to have fun when I handed them their exams. Your first reaction to that might be similar to their reactions—"Wise a**!" However, I hope that after reading this book, you will come to understand—as my students did—learning Calculus II is a chance to show yourself what you can do when you set your mind to it and have fun with this book. You never know what you can do until you try.

How This Book Is Organized

This book is presented in four parts:

In **Part 1, Review of Pre-Calculus and Calculus I,** you review a few key topics that you covered in Pre-Calculus and come into play in Calculus II. You'll also do a quick review of limits and derivatives from Calculus I.

In **Part 2, Length, Area, and Volumes,** you learn about the applications of integration to compute one-, two-, and three-dimensional measurements.

In **Part 3, More Definite and Indefinite Integrals,** you study integration techniques beyond the notion of the simple antiderivative.

In **Part 4, The Infinite Series and More,** you look at topics that extend the notions of the Rectangular Coordinate System as well as topics to extend what you have learned about integration. Finally, you end with a study of topics that are very applicable to how your calculator does math. I've also included a final exam as the last chapter so you can assess your understanding of what you've learned.

At the very end of the book, I've included solutions to the You've Got Problems sidebars throughout the book. Also, you'll find an appendix that provides you more integration practice as well as a glossary of helpful terms.

Extras

Throughout the book, you'll see helpful sidebars that reinforce what you're learning. Here's what to look for:

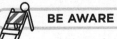 **CRITICAL POINT**

These sidebars are meant to draw your attention to key issues in calculus and key people who have been very influential in the development of the topic.

DEFINITION

These sidebars help you break down the terms used in calculus so you can better understand what is presented to you.

BE AWARE

Although I warn you about common pitfalls and dangers throughout the book, these sidebars deserve special attention. They are also meant to draw a special light on critical errors students often make.

 YOU'VE GOT PROBLEMS

After I discuss a topic, I explain how to work out a certain type of problem and then you get to try it on your own. These problems are very similar to those I walk you through in the chapters, but now it's your turn to shine. Even though all the answers appear in Appendix A, you should only look there to check your work.

Acknowledgments

I want to thank the thousands of students who were in my classes over the years, especially the students from BC Calculus at Niskayuna High School in Niskayuna, New York. You did a great job of preparing me to write this book.

Special Thanks to the Technical Reviewer

Idiot's Guides: Calculus II was reviewed by an expert who double-checked the accuracy of what's presented here to help us ensure learning Calculus II is as easy as it gets. Special thanks are extended to Angela Netoskie.

Angela is a mathematics teacher at Niskayuna High School, Niskayuna, New York. She has taught Algebra, Common Core Algebra, Geometry, Algebra 2/Trigonometry, Pre-Calculus, Accelerated Pre-Calculus, Calculus, Advanced Placement AB Calculus, and Advanced Placement BC Calculus. She holds a Master of Arts in mathematics from the State University of New York at Albany, and a Bachelor of Science from Pace University.

Review of Pre-Calculus and Calculus I

The first part of this book is intended to refresh the work you did in Calculus I and review some topics you might have studied in your Pre-Calculus class that are applied in Calculus II.

In addition, we review trigonometry and logarithms—topics you covered in Calculus I.

I also include material on parametric equations, the polar coordinate system, and partial fractions.

Pre-Calculus Topics Used in Calculus II

Pre-Calculus is usually the last class a student takes before entering Calculus I. The topics taught in Pre-Calculus vary from state to state, school to school, and sometimes, from teacher to teacher. There are a few topics which often form the mainstay of the course simply because they are key building blocks to one's ability to answer topics in Calculus (as opposed to "understanding" calculus). The topics included in this chapter are the ones my students needed to review as we were doing the calculus.

In This Chapter

- Reviewing key trigonometric relationships used in calculus

- Euler's number and his logarithms

- The other coordinate system for the plane

- Understanding x does not always have to be the independent variable

- Learning still more about fractions

Trigonometry

The unit of angle measurement in higher mathematics is the radian. Most students in the United States and Canada do not encounter radian value until they study trigonometric functions. The right triangle trigonometry uses in geometry can be done with degrees, but the applications to periodic phenomena, such as tides and alternating current, require the use of radians because the radian is a unitless entity.

For example, if you drive 300 miles in 5 hours, your average speed is 60 miles per hour. Average speed has units. The radian measure for an angle is computed by measuring the length of an arc of a circle and dividing it by the length of the radius of that circle. An arc of a circle with radius 4 inches that contains one sixth of the entire circumference of the circle has a length of $\frac{1}{6}(2\pi \times 4) = \frac{4\pi}{3}$ inches. The measure of the angle that forms this arc has measure of $\frac{\frac{4\pi}{3}}{4} = \frac{\pi}{3}$ radians. It is beneficial for you to know that 180 degrees corresponds to π radians. You can use proportions to find the corresponding radian values for familiar degree measurements.

YOU'VE GOT PROBLEMS

Problem 1: How many radians correspond to 30 degrees?

Angle measurements may be positive or negative. A positive angle measurement corresponds to a counterclockwise rotation from what we can call the positive side of the x-axis and a negative angle measurement corresponds to a clockwise rotation. Therefore, an angle with measure $\frac{5\pi}{4}$ radians terminates at the same place as an angle with measure $\frac{-3\pi}{4}$ radians.

CRITICAL POINT

Why are there 360 degrees in a circle? Why not 100 degrees, as in 100 percent of the circle? Blame it on the Babylonians. Unlike our decimal system, the Babylonian number system was based on 60. Sixty has a lot of factors (1, 2, 3, 4, 5, 6, 10, 12, 15, 20, 30, 60) so subdividing quantities into fractions was easier to do. Their year was divided into 6 periods of 60 days, or 360 days to come "full circle." They compensated for the missing 5 days with a religious festival after the harvest so their calendar was fairly consistent. The issue of leap years never became a problem because the civilization didn't last long enough for their calendar to be noticeably out of sync with Earth's orbit.

The three basic trigonometric functions are the sine, tangent, and secant, which are traditionally abbreviated to sin, tan, and sec. (Are you surprised not to see cosine?) The remaining three trigonometric functions, the co-functions, are cosine, cotangent, and cosecant (cos, cot, and csc, respectively). The relationship between the three basic functions and the co-functions is a phase shift. That is:

$$\sin(x) = \cos\left(\tfrac{\pi}{2} - x\right) \qquad \tan(x) = \cot\left(\tfrac{\pi}{2} - x\right) \qquad \sec(x) = \csc\left(\tfrac{\pi}{2} - x\right)$$

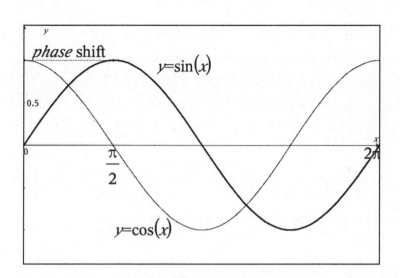

Figure 1.1

The graphs of y = sin(x) and y = cos(x) are horizontal translations of one another. The amount of the translation is called the phase shift.
These functions are also related as:

$$\tan(x) = \frac{\sin(x)}{\cos(x)} \qquad \sec(x) = \frac{1}{\cos(x)} \qquad \csc(x) = \frac{1}{\sin(x)}$$

There are three Pythagorean identities involving these functions (based on the Pythagorean theorem for right triangles):

$$\sin^2(x) + \cos^2(x) = 1 \qquad 1 + \tan^2(x) = \sec^2(x) \qquad 1 + \cot^2(x) = \csc^2(x)$$

The double angle identities for sine and cosine are:

$$\sin(2x) = 2 \sin(x) \cos(x)$$

$$\cos(2x) = \cos^2(x) - \sin^2(x)$$

$$= 1 - 2 \sin^2(x)$$

$$= 2 \cos^2(x) - 1$$

Example 1: If $\cos(x) = A$ and $\sin(x) > 0$, write expressions for $\sin(2x)$ and $\cos(2x)$.

Solution: Use the Pythagorean identity $\sin^2(x) + \cos^2(x) = 1$ to find the value of $\sin(x)$:

$$A^2 + \sin^2(x) = 1 \text{ becomes } \sin^2(x) = 1 - A^2 \text{ and } \sin(x) = \sqrt{1 - A^2}$$

$$\sin(2x) = 2\sin(x) \cos(x) = 2A\sqrt{1 - A^2}$$

$$\cos(2x) = 2\cos^2(x) - 1 = 2A^2 - 1$$

Each of the trigonometric functions has inverses, but the two most important (and frequently used) are the inverse sine (arcsin or $\sin^{-1}$) and the inverse tangent (arctan or $\tan^{-1}$).

The domain and range of these functions are as follows:

	Domain	**Range**
$y = \sin^{-1}(x)$	$[-1, 1]$	$\left[\frac{-\pi}{2}, \frac{\pi}{2}\right]$
$y = \tan^{-1}(x)$	$(-\infty, \infty)$	$\left(\frac{-\pi}{2}, \frac{\pi}{2}\right)$

YOU'VE GOT PROBLEMS

Problem 2: If $\sin(A) = \frac{1}{\sqrt{10}}$ and $\cos(A) < 0$, determine the value for $\sin(2A)$.

Exponents and Logarithms

Almost every junior high school student and high school student will tell you that the value of π is approximately 3.14 (though they rarely say approximately). You should develop a comfort level with the number e, Euler's number, being approximately 2.718. Euler's number is the most frequently used base for exponential functions and is used as the base of the natural logarithms, written ln. The value of e is determined by evaluating the limit:

$$\lim_{n \to \infty} \sum_{k=0}^{n} \frac{1}{k!} = \lim_{n \to \infty} \left(1 + 1 + \frac{1}{2!} + \frac{1}{3!} + \dots + \frac{1}{n!}\right)$$

CRITICAL POINT

Late in the 1990s when everyone seemed to be driven by Y2K mania, a group of professional mathematicians were asked who were the most influential mathematicians of the millennium. Isaac Newton edged out Carl Gauss by a very narrow margin. Euler was third on that list, not too far behind them.

The domain of $f(x) = e^x$ is the set of real numbers while the range is $y > 0$. If $e^a = b$ then it is the case that $ln(b) = a$ because $f(x) = e^x$ and $g(x) = \ln(x)$ are inverse functions. An important property of the inverse functions $f(x)$ and $g(x)$ is that $f(g(x)) = g(f(x)) = x$. In this case, we get $e^{\ln(x)} = x$ and $ln(e^x) = x$.

Three other properties of exponential functions, and their corresponding impact on the logarithmic functions, are as follows:

$$e^{x+y} = e^x \times e^y \qquad\qquad \ln(xy) = \ln(x) + \ln(y)$$

$$e^{x-y} = \frac{e^x}{e^y} \qquad\qquad\qquad \ln\left(\frac{x}{y}\right) = \ln\left(x\right) - \ln\left(y\right)$$

$$(e^x)^y = e^{xy} \qquad\qquad\qquad ln(x^y) = y\ln(x)$$

If you recall that a logarithm is an exponent, then the rules make a bit more sense. The rule …

$$ln(xy) = ln(x) + ln(y)$$

… says to find the exponent for the product, you add the exponents for the terms involved.

Let's try a problem that is a little more complicated.

Example 2: Simplify $\ln\left(\frac{(x+3)^2}{(x-4)(x+1)}\right)$.

Solution: Use the second rule for quotients to get $\ln\left(\frac{(x+3)^2}{(x-4)(x+1)}\right) = \ln((x+3)^2) - \ln((x-4)(x+3))$.

Use the third rule for powers:

$$2\ ln(x+3) - ln((x-4)(x+1))$$

Use the first rule for products (but be careful with the subtraction sign in front of this term):

$$2\ ln(x+3) - (ln(x-4) + ln(x+1))$$

Distribute the subtraction sign:

$$2\ ln(x+3) - ln(x-4) - ln(x+1)$$

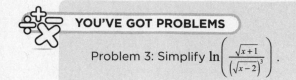

Problem 3: Simplify $\ln\left(\dfrac{\sqrt{x+1}}{\left(\sqrt{x-2}\right)^3}\right)$.

Parametric Equations

You can write the vast majority of functions in the form y is a function of x. You can write equations in x and y for relations that are not functions such as circles, ellipses, some hyperbolas, and parabolas. A *parametric equation* uses a third variable, or parameter, as the independent variable and has x and y depend on this parameter.

📖 **DEFINITION**

A **parametric equation** is one in which both x and y are functions of a third variable.

Example 3: What is the difference among the three parametric equations?

$x = \cos(t)$ $\qquad\qquad$ $x = \cos(5t)$ $\qquad\qquad$ $x = \sin(3t)$

$y = \sin(t)$ $\qquad\qquad$ $y = \sin(5t)$ $\qquad\qquad$ $y = \cos(3t)$

Solution: The first two equations create a circle drawn in a counterclockwise direction with the second graph being drawn faster than the first. The third equation draws the circle in a clockwise motion. Put your graphing calculator in parametric mode and verify the results.

Notice that all three equations take advantage of the Pythagorean identity $\sin^2(x) + \cos^2(x) = 1$.

Example 4: What is the difference between the graphs of the parametric equations?

$x = t$ and $x = \sqrt{t}$

$y = t^2$ and $y = t$

Solution: Put your graphing calculator in parametric mode and sketch the first set of equations to see the parabola. However, the second set of equations only draws the right-hand side of the parabola.

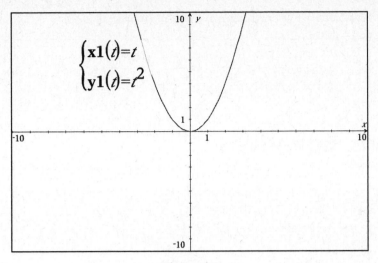

Figure 1.2

A parabola defined parametrically.

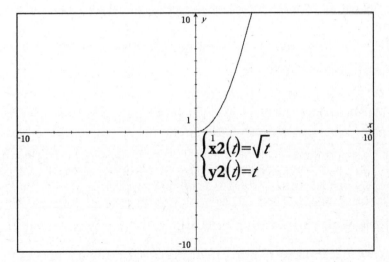

Figure 1.3

The right half of a parabola defined parametrically.

You can also make some very interesting pictures in parametric mode:

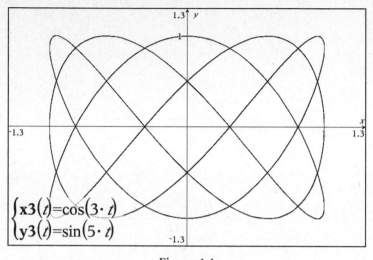

Figure 1.4
The parametric equation x = cos(3t), y = sin(5t).

Polar Coordinates

The branch of mathematics called coordinate geometry began when Rene Descartes was lying in his bed and realized he could divide the ceiling with a series of vertical and horizontal lines. Location on the ceiling could be determined by counting the number of vertical and horizontal lines from some fixed point. You and I have been using the *Cartesian* (rectangular) *Coordinate plane* our entire lives. An immediate impact of this system is that each point has a unique set of coordinates and each set of coordinates corresponds to a unique point.

DEFINITION

A **polar coordinate plane** is a plane in which location from a central point is defined by the length of the radius of a circle drawn from that point and the measure of an angle drawn from a fixed ray using the central point as its endpoint. The **Cartesian Coordinate plane** is formed from a central point with two perpendicular lines passing through this point. Coordinates are determined by the distance you move to the right (positive designation) or to the left (negative) and then up (positive) or down (negative).

However, what if Descartes had done it differently? What if he marked a central point on the ceiling as his fixed point, drew a series of concentric circles from this point, and drew a ray (called a pole) from this central point for which he could draw angles? He could then describe the location of a point on the ceiling by indicating the radius of a particular circle and also the angle of the ray from the central point to the point in question made with the pole. Every point on the plane of the ceiling can be identified. This type of a coordinate system is called a *polar coordinate plane.*

CRITICAL POINT

A major difference between this scheme and the Cartesian Coordinate system is while each set of coordinates describes a unique point, each point does not have a unique set of coordinates. For example, the point with coordinates $\left(4, \frac{5\pi}{4}\right)$ is also the point with coordinates $\left(4, \frac{-3\pi}{4}\right)$, $\left(4, \frac{-11\pi}{4}\right)$, and $\left(4, \frac{13\pi}{4}\right)$, to name a few.

In the accompanying diagram, point P has Cartesian coordinates (x,y) and polar coordinates (r,θ). To convert between the two coordinate systems, you can use right triangle trigonometry.

$$x = r\cos(\theta) \qquad x^2 + y^2 = r^2$$

$$y = r\sin(\theta) \qquad \theta = \tan^{-1}\left(\frac{y}{x}\right)$$

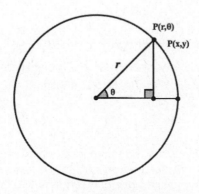

Figure 1.5

A point P in a plane with its rectangular coordinates (x, y) and its polar coordinates (r, θ).

Here is a problem similar to the problem you just did but with a twist to it.

Example 5: Convert the rectangular coordinates $\left(-4, 4\sqrt{3}\right)$ to polar coordinates.

Solution: Find the value of r: $\left(-4\right)^2 + \left(4\sqrt{3}\right)^2 = r^2$ becomes $16 + 48 = 64 = r^2$ so $r = 8$.

We need to be careful about the value of θ because the point $\left(-4, 4\sqrt{3}\right)$ is in the second quadrant implying that the primary value for θ should be between $\frac{\pi}{2}$ and π. The value for $\tan^{-1}\left(\frac{4\sqrt{3}}{-4}\right)$ is -1.0472 if you are using a non-CAS calculator or $\frac{-\pi}{3}$ (if you remember all the values from the unit circle from your days in trigonometry or you have a CAS operating system on your graphing calculator). (A CAS system has the ability to give symbolic answers as well as numerical.) You must add π to this reference angle to get the correct angle. The correct polar coordinates are $\left(8, \frac{2\pi}{3}\right)$ in terms of π or $(8, 2.094395)$ numerically.

Graphs of functions look much different in polar coordinates than they do in rectangular coordinates. For example:

- The graph of $r = \sin(\theta)$ is a circle that is tangent to the pole at the origin.

- The graph of $r = 4$ is also a circle centered at the origin.

- The graph of $\theta = 2$ is a line.

- The graph of $r = \theta$ is a spiral.

Try graphing these on your graphing utility to verify this.

Three of the more important polar graphs that come into play in Calculus II are as follows:

- The rose ($r = a \sin(n\theta)$ or $r = a \cos(n\theta)$)

- The cardioid ($r = a + a \sin(\theta)$ or $r = a + a \cos(\theta)$)

- The limacon ($r = a + b \sin(\theta)$ or $r = a + b \cos(\theta)$)

You might be asked to compute the slope of a tangent line to a curve at a particular point or to find the area under one of the curves. It would be worth your while to use your graphing calculator (or to use your web browser to find an interactive graphing calculator site) to play with some of these graphs. The traditional rules have the parameters n, a, and b be integers and have the domain for θ be $[0, 2\pi]$.

Example 6: For what values of θ does the graph of $r = 3\cos(2\theta)$ pass through the center of the coordinate system?

Solution: The graph of $r = 3\cos(2\theta)$ is the 4 petal rose as shown. Point F has coordinates (3, 0) so the first plotted point sits on the pole. As the value of θ increases, point F will move along the top of the first petal.

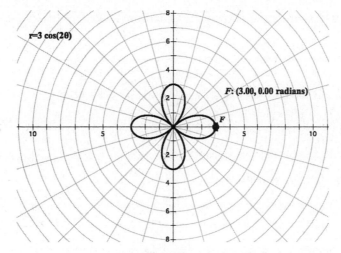

Figure 1.6

The graph of the polar rose with equation $r = 3\cos(2\theta)$ with the point (3, 0) indicated.

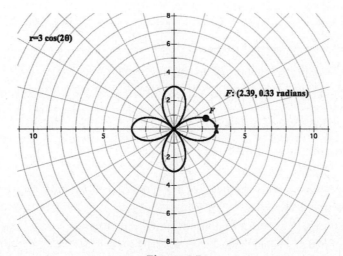

Figure 1.7

As θ increases from 0 to 0.33 radians, the point travels along one of the petals of the rose.

The question becomes "For what value of θ does cos(2θ) first equal 0?" We know that $\cos\left(\frac{\pi}{2}\right) = 0$ so we set $2\theta = \frac{\pi}{2}$ to get $\theta = \frac{\pi}{4}$, or $\theta = 0.7854$.

YOU'VE GOT PROBLEMS

Problem 4: Convert the rectangular coordinates $\left(5, 5\sqrt{3}\right)$ to polar coordinates.

Geometric Sequences and Series

A *geometric sequence* is a list of numbers in which consecutive terms have a common ratio. For example, each term in the sequence 1, 2, 4, 8, 16, 32, ... is found by multiplying the previous term by 2. In the sequence 2187, 1458, 972, 648, 432, ... each term is $\frac{2}{3}$ of the previous term. In these cases, 2 and $\frac{2}{3}$ are the common ratios.

Example 7: What is the common ratio for each of the following sequences?

1. 1, 2, 4, 8, 16, 32, ...

2. 128000, 64000, 32000, 16000, ...

3. $3, 3\sqrt{2}, 6, 6\sqrt{2}, 12, ...$

Solution: Each term in sequence 1 is twice as large as the preceding term. The common ratio is 2.

In sequence 2, each term is half as large as the preceding term. The common ratio is $\frac{1}{2}$.

In sequence 3, the ratios $\frac{3\sqrt{2}}{3}, \frac{6}{3\sqrt{2}}, \frac{6\sqrt{2}}{6},$ and $\frac{12}{6\sqrt{2}}$ are all $\sqrt{2}$, making $\sqrt{2}$ the common ratio.

A *geometric series* is the sum of the terms of a geometric sequence. The formula for computing the sum S of the first n terms of a geometric series with first term a and common ratio r is $S = \frac{a\left(1-r^n\right)}{1-r}$. This formula has a lot of nice applications, particularly in the world of finance.

However, it does not have much application in Calculus II, with one exception. What if the number of terms is infinitely large?

The series corresponding to sequence 1 in Example 7 is $1 + 2 + 4 + 8 + 16 + 32 + \ldots$ The sum of an infinite number of terms for this series would be $S = \frac{1(1-2^{\infty})}{1-2}$. What is the value of 2 raised to an infinitely large power? It is an infinitely large number. Therefore, the sum of the terms in series 1 would grow infinitely large (that is, it diverges).

The series corresponding to sequence 2 will have a sum $S = \frac{128000\left(1-\left(\frac{1}{2}\right)^{\infty}\right)}{1-\frac{1}{2}}$. The value of $\left(\frac{1}{2}\right)^{\infty}$ will go to 0 because the denominator of the fraction grows infinitely large. Therefore, series 2 converges to the sum $S = \frac{128000(1-0)}{\frac{1}{2}} = 256000$.

Can you see that the series corresponding to sequence 3 will diverge? Because $\sqrt{2} > 1$, $\left(\sqrt{2}\right)^{\infty}$ will grow infinitely large.

YOU'VE GOT PROBLEMS

Problem 5: Determine the sum of the series $12 + 8 + \frac{16}{3} + \frac{32}{9} + \ldots$.

Partial Fractions

You learned how to add fractions in elementary school and how to add algebraic fractions in Algebra I. Have you ever had to reverse the process? That is, have you tried to determine what two fractions when added yield $\frac{7}{12}$.

Two quick answers might be $\frac{1}{12}, \frac{6}{12} = \frac{1}{2}$ and $\frac{1}{3}, \frac{1}{4}$.

The procedure for separating algebraic fractions—something that you will need to do in order to perform certain integration problems—is called *partial fractions*.

DEFINITION

A **partial fraction** is each of the two or more fractions that can be added to form a more complex fraction.

Example 8: Separate the algebraic fraction $\frac{8x+1}{x^2-x-6}$ into the two fractions that are its addends (the values that are added together).

Solution: First factor the denominator into its prime factors, $(x-3)(x+2)$. Each of these factors are linear expressions, so we can rewrite the original fraction with each of these factors as denominators and constants as numerators. That is, $\frac{8x+1}{x^2-x-6} = \frac{A}{x-3} + \frac{B}{x+2}$.

The problem is to now determine the values of A and B. The process is fairly straightforward. Multiply both sides of the equation by the common denominator, $(x-3)(x+2)$:

$$8x + 1 = A(x + 2) + B(x - 3)$$

Let $x = 3$, causing the term $B(x - 3)$ to become 0:

$$8(3) + 1 = A(3 + 2)$$

$$25 = 5A, \text{ so } A = 5$$

Now let $x = -2$ so $A(x + 2)$ becomes 0:

$$8(-2) + 1 = B(-2 - 3)$$

$$-15 = -5B, \text{ so } B = 3$$

Therefore, $\frac{8x+1}{x^2-x-6} = \frac{5}{x-3} + \frac{3}{x+2}$.

When decomposing the algebraic fraction into its component parts, the degree of the numerator will always be one less than the degree of the denominator. So when one of the denominators is a quadratic factor that cannot be factored, the numerator will be in the form $Ax + B$.

Example 9: Decompose $\frac{3x^2+4x+15}{(x^2+1)(x+3)}$.

Solution: The factors of the denominator are given to you. The term $x^2 + 1$ is quadratic so the corresponding numerator will be linear. Therefore, $\frac{3x^2+4x+15}{(x^2+1)(x+3)}$ will be rewritten as $\frac{3x^2+4x+15}{(x^2+1)(x+3)} = \frac{Ax+B}{x^2+1} + \frac{C}{x+3}$.

As we did before, multiply through by the common denominator:

$$3x^2 + 4x + 15 = (Ax + B)(x + 3) + C(x^2 + 1)$$

Let $x = -3$ to eliminate the term $(Ax + B)(x + 3)$:

$3(-3)^2 + 4(-3) + 15 = C((-3)^2 + 1)$

$30 = 10C$, so $C = 3$

Unlike the earlier problem, we will not be able to eliminate terms by picking convenient values for x to quickly get the values of A and B. However, by choosing $x = 0$, the term Ax can be eliminated.

Let $x = 0$:

$15 = B(0 + 3) + 3((0)^2 + 1)$

$15 = 3B + 3$, so $B = 4$

To find the value of A, pick any number that you would like to use. My thought is to keep things simple, so I will choose $x = 1$:

$3(1)^2 + 4(1) + 15 = (A(1) + 4)(1 + 3) + 3((1)^2 + 1)$

$22 = (A + 4)(4) + 3(2)$

$22 = 4A + 16 + 6$

$0 = 4A$, so $A = 0$

Therefore, $\dfrac{3x^2 + 4x + 15}{(x^2+1)(x+3)} = \dfrac{4}{x^2+1} + \dfrac{3}{x+3}$.

YOU'VE GOT PROBLEMS

Problem 6: Separate the algebraic fraction $\dfrac{x-19}{x^2+4x-5}$ into the two fractions that are its addends.

The Least You Need to Know

- The Pythagorean identities and double angle identities will have a lot of applications in Calculus II.
- The independent variable of a function isn't always x; sometimes it is t.
- When a number larger than 1 is raised to infinity, the result is infinitely large. When a number between 0 and 1 is raised to infinity, the result is 0.
- You will still be using some of your basic algebra skills (for example, factoring, solving systems of equations, and simplifying fractions) in Calculus II.

CHAPTER 2

Limits, Derivatives, and Basic Integration

Calculus is essentially the study of two processes, instantaneous rates of change and accumulation. To study the notion of instantaneous rates of change, you first need to develop the concept of a "mathematical microscope," the limit. Do you remember when you were first learning to graph a line? You were directed to make a table of values (at least three points to verify that you did not make an arithmetic mistake), plot those points on a grid and then use a straight edge to draw the line through the plotted points. Did it ever occur to you that maybe the "line" you were drawing was actually perforated like a sheet of paper that you can rip out of your notebook? I didn't either when I was that age. Now that you are studying calculus, you must prepare to deal with a large variety of functions, not just the "nice" functions that you saw in algebra.

In This Chapter

- Understanding the notion of a limit and continuity
- Working with the derivative and differentiation formulas
- Implicit differentiation
- Solving max-min problems
- Computing related rates problems

Consider the function $f(x) = \begin{cases} 1 & \text{, if } x \text{ is rational} \\ -1 & \text{, if } x \text{ is irrational} \end{cases}$. The graph of this function appears to be two horizontal lines, one at $y = 1$ and the other at $y = -1$. However, closer inspection shows that these lines are indeed perforated because between two rational numbers, say 0.98 and 0.99, there are an infinite number of irrational numbers. The purpose of the example is to remind you that when a mathematician writes a theorem about a function, the theorem applies to all functions, not just the "nice" ones, those that you have studied up to this point in your mathematical career.

Limits

For a limit to exist, it is necessary that whenever an interval is established around an input value, no matter how small, there exists a corresponding interval around a unique value of the output. (These intervals do not have to be the same size.)

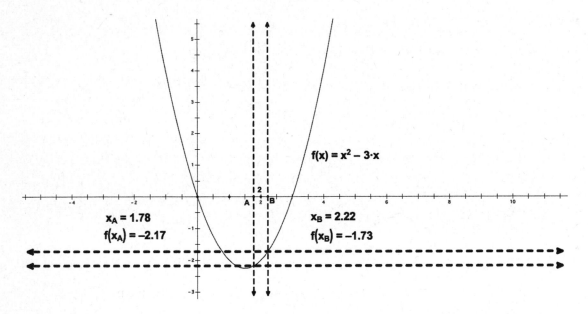

Figure 2.1
Parabola with limit.

In Figure 2.1, points *A* and *B* are symmetric about $x = 2$. The thick dashed horizontal lines represent their functional values. Regardless of the intervals size from which $x = 2$ is formed, there will always be a corresponding interval of $y = -2$. Therefore, we conclude that as x approaches 2, f(x) approaches -2. Written symbolically, $\lim_{x \to 2} x^2 - 3x = -2$.

If g(x) is the *piecewise function* defined by $g(x) = \begin{cases} -3x + 4 & x < 1 \\ 2 & x = 1 \\ x^2 & x > 1 \end{cases}$. What is $\lim_{x \to 1} g(x)$?

DEFINITION

A **piecewise function** is a function that's defined by a set of expressions, each with its own domain.

As the graph shows, as the values of x get very near to 1, the values of g(x) get very near to 1.

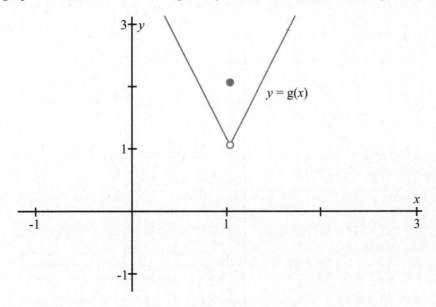

Figure 2.2
The limit of the function as x approaches 1 is not the same as the value of the function at x = 1.

It is true that when $x = 1$, $g(x) = 2$. What happens at a value of the input is the value of the function. What happens in the neighborhood of the input value is the limit of the function. This is a very fine point, but one that is critical to your understanding of limits. You have the functional value $g(1) = 2$ and the limiting value $\lim\limits_{x \to 1} g(x) = 1$.

A function f(x) is said to be *continuous at a point* $x = c$ if and only if (abbreviated iff) $\lim\limits_{x \to c} f(x) =$ f(c) and f(c) exists. A function is said to be continuous if it is continuous at all points in the domain of the function. In essence, a continuous function is one that you can draw by putting your pencil point on the paper and never having to lift the point off the paper to complete the

graph. The function f(x) = x^2 – 3x is continuous while the function $g(x) = \begin{cases} -3x + 4 & x < 1 \\ 2 & x = 1 \\ x^2 & x > 1 \end{cases}$ is

not continuous at $x = 1$.

DEFINITION

A function is **continuous at a point** $x = c$ iff $\lim\limits_{x \to c} f(x) =$ f(c). A function is continuous if it is continuous at all points.

Example 1: Determine $\lim\limits_{x \to 3} k(x)$ if $k(x) = \begin{cases} 2x - 1 & x < 3 \\ x^2 - 3 & x \geq 3 \end{cases}$.

Solution: Because the function is defined with split domains, use a left-hand limit for $x < 3$ and a right-hand limit for $x \geq 3$.

You get $\lim\limits_{x \to 3^-} 2x - 1 = 5$ and $\lim\limits_{x \to 3^+} x^2 - 3 = 6$. These two values are not the same, indicating that a single value of the output cannot be isolated, so we say $\lim\limits_{x \to 3} k(x)$ does not exist (abbreviated DNE).

Example 2: Evaluate $\lim\limits_{x \to 2} \frac{x^2 - x - 2}{x^2 - 4}$.

Solution: Substitute 2 for x and evaluate the expression to get $\frac{0}{0}$. When you get this kind of answer, we call the result *indeterminate,* because we cannot decide the value of the limit. However, by substituting 2 for x and getting a result of 0, you know $x - 2$ is a factor of the expression.

Rewrite $\frac{x^2 - x - 2}{x^2 - 4}$ as $\frac{(x-2)(x+1)}{(x-2)(x+2)}$ and then reduce the fraction to $\frac{x+1}{x+2}$. Now evaluate the limit $\lim\limits_{x \to 2} \frac{x+1}{x+2}$ to get the answer $\frac{3}{4}$.

To evaluate limits as the input goes to positive or negative infinity, take advantage of the fact that the largest term in any polynomial determines end behavior and that $\lim\limits_{x \to \infty} \frac{1}{x} = 0$.

Example 3: Evaluate $\lim\limits_{x \to \infty} \dfrac{4x^3 - 5x - 3}{3x^3 + 5x^2 - 100x}$.

Solution: You can physically (or mentally) divide the numerator and denominator by x^3 to get $\lim\limits_{x \to \infty} \dfrac{4 - \frac{5}{x^2} - \frac{3}{x^3}}{3 + \frac{5}{x} - \frac{100}{x^2}}$. As x approaches infinity, all terms in x go to 0 so the value of the limit is $\frac{4}{3}$.

✏️ **CRITICAL POINT**

Two important limits that appear at different points in the curriculum are
$$\lim_{x \to 0} \frac{\sin(x)}{x} = 1 \text{ and } \lim_{x \to \infty} \left(1 + \frac{1}{x}\right)^x = e .$$

The last issue on limits we need to discuss in this review involves limits that go to infinity—not the case in which $x \to \infty$ but the cases in which $y \to \infty$ What is the difference between $\lim\limits_{x \to 0} \frac{1}{x}$ and $\lim\limits_{x \to 0} \frac{1}{x^2}$? In the case of $f(x) = \frac{1}{x}$, as x approaches 0 from the negative side of the x-axis, the graph goes toward negative infinity. However, as x approaches 0 from the positive side of the x-axis, the graph goes toward positive infinity. These two different results make it fairly easy to understand that $\lim\limits_{x \to 0} \frac{1}{x} = $ DNE. In the case of $g(x) = \frac{1}{x^2}$, whether x approaches 0 from the left or the right, the graph will approach positive infinity. In this case, we say $\lim\limits_{x \to 0} \frac{1}{x^2} = \infty$. Even though we cannot isolate a single value, we describe the behavior of the function by indicating that both branches of the graph are going in the same direction.

✈️ **YOU'VE GOT PROBLEMS**

Problem 1: Evaluate $\lim\limits_{x \to -2} \dfrac{2x^2 - x - 10}{6 + x - x^2}$.

Derivatives

Contrary to what many believe, Isaac Newton was never hit on the head with an apple. However, after he did see an apple fall in the orchard, he realized he had the means to discuss instantaneous rates of change. (Remember, Newton was trying to determine what gravity was and much of his work involved describing the effects of gravity.)

Newton knew he could measure the distance that the apple fell and the time it took for the apple to fall. His reasoning went something like this:

1. Suppose I want to know the velocity the apple is traveling when it is 6 feet off the ground.

2. I can measure the time the apple takes to move from 8 feet to 4 feet off the ground and compute the average velocity during that time span.

3. Then I can repeat the process as the apple falls from 7 feet to 5 feet, 6.5 feet to 5.5 feet, 6.1 feet to 5.9 feet, and so on.

4. I will have a sequence of average velocities over smaller and smaller intervals, and the limit of this sequence will be the instantaneous velocity.

If f(t) represents the number of feet above ground that the apple is located then $\lim\limits_{t \to a} \dfrac{f(t) - f(a)}{t - a}$

represents the limit of the average velocities, or the instantaneous velocity. This limit is known as the derivative and the symbol for it is f'(x).

CRITICAL POINT

Newton actually called the change a fluxion and used a dot above the f rather than a prime next to the f to symbolize this. Changes in name and notation were made by others later.

Other representations for the definition of the derivative are $\lim\limits_{\Delta x \to 0} \dfrac{f(x + \Delta x) - f(x)}{\Delta x}$ and $\lim\limits_{h \to 0} \dfrac{f(x + h) - f(x)}{h}$.

(It is suspected that the use of h rather than delta x was caused by publishers so they could save on the cost of ink.)

The formulas for differentiation can all be derived from the definition. The rules for differentiation are as follows:

- Constant Rule: If f(x) = cg(x) where c is a constant, then f'(x) = cg'(x).

- Sum Rule: If f(x) = g(x) + k(x), then f'(x) = g'(x) + k'(x).

- Product Rule: If f(x) = g(x) × k(x), then f'(x) = g'(x) × k(x) + g(x) × k'(x).

- Quotient Rule: If $f(x) = \dfrac{g(x)}{k(x)}$ then $f'(x) = \dfrac{g'(x) k(x) - g(x) k'(x)}{(k(x))^2}$.

- Chain Rule: If f(x) = g(k(x)) then f'(x) = g'(k(x)) × k'(x).

The following table shows the most-used functions and their derivatives.

Function	Derivative	Function	Derivative
$f(x) = x^n$	$f'(x) = n\, x^{n-1}$	$f(x) = \cos(x)$	$f'(x) = -\sin(x)$
$f(x) = e^x$	$f'(x) = e^x$	$f(x) = \tan(x)$	$f'(x) = \sec^2(x)$
$f(x) = b^x$	$f'(x) = \ln(b)\, b^x$	$f(x) = \sec(x)$	$f'(x) = \sec(x)\tan(x)$
$f(x) = \ln(x)$	$f'(x) = \dfrac{1}{x}$	$f(x) = \csc(x)$	$f'(x) = -\csc(x)\cot(x)$
$f(x) = \log_b(x)$	$f'(x) = \dfrac{1}{x\ln(b)}$	$f(x) = \cot(x)$	$f'(x) = -\csc^2(x)$
$f(x) = \sin(x)$	$f'(x) = \cos(x)$	$f(x) = \sin^{-1}(x)$	$f'(x) = \dfrac{1}{\sqrt{1-x^2}}$
		$f(x) = \tan^{-1}(x)$	$f'(x) = \dfrac{1}{1+x^2}$

Example 4: Find the derivative of $f(x) = \ln(\sec(x))$.

Solution: Because this is a composite function, apply the Chain Rule to get $f'(x) =$
$\left(\frac{1}{\sec(x)}\right)\left(\sec(x)\tan(x)\right) = \tan(x)$.

Example 5: Find the derivative of $g(x) = \tan^{-1}\left(\sqrt{x-1}\right)$.

Solution: Apply the Chain Rule to get $g'(x) = \left(\dfrac{1}{1+\left(\sqrt{x-1}\right)^2}\right)\left(\frac{1}{2}(x-1)^{-\frac{1}{2}}\right) = \left(\dfrac{1}{1+x-1}\right)\left(\dfrac{1}{2\sqrt{x-1}}\right) = \dfrac{1}{2x\sqrt{x-1}}$.

CRITICAL POINT

England was often at odds with, if not at war with, many of the countries on mainland Europe during the sixteenth through eighteenth centuries. That animosity carried over to the development of calculus. Isaac Newton founded the basis of calculus in England, but it was a German mathematician, Gottfried Leibniz, who first published materials about calculus. Newton used a variation on the prime notation, $f'(x)$, to represent the derivative, but Leibniz preferred the notion that the derivative was a measure of slope, so he extended the notion of $\frac{\Delta y}{\Delta x}$ for the slope of a line to $\frac{dy}{dx}$ for the slope of the tangent line. Both notations have made their way through history and are used interchangeably.

Example 6: Find the value of $\frac{dy}{dx}$ when $x = 4$ if $y = \frac{x^2 \sqrt[3]{5x-12}}{e^{x-4}}$.

Solution: Although this problem can be solved using a combination of the Quotient, Product, and Chain Rules, use the technique of *logarithmic differentiation* to simplify the problem:

$$\ln(y) = \ln\left(\frac{x^2 \sqrt[3]{5x-12}}{e^{x-4}}\right) = \ln\left(x^2\right) + \ln\left(\sqrt[3]{5x-12}\right) - \ln\left(e^{x-4}\right) =$$
$$2\ln(x) + \tfrac{1}{3}\ln(5x-12) - (x-4)$$

Now you can take the derivative of both sides (remember, y is a function of x so the Chain Rule applies) …

$$\tfrac{1}{y}\tfrac{dy}{dx} = 2\left(\tfrac{1}{x}\right) + \tfrac{1}{3}\left(\tfrac{1}{5x-12}\right)(5) - 1$$

… which becomes $\frac{dy}{dx} = y\left(2\left(\tfrac{1}{x}\right) + \tfrac{1}{3}\left(\tfrac{5}{5x-12}\right) - 1\right) = \frac{x^2 \sqrt[3]{5x-12}}{e^{x-4}}\left(\tfrac{2}{x} + \tfrac{1}{3}\left(\tfrac{5}{5x-12}\right) - 1\right).$

At $x = 4$, the value of the derivative is $\frac{16\sqrt[3]{8}}{e^0}\left(\tfrac{1}{2} + \tfrac{5}{24} - 1\right) = \frac{-28}{3}$.

YOU'VE GOT PROBLEMS

Problem 2: Find the derivative of the function p(x) = x ln(x) − x.

Example 7: Given $y = x \sin(x)$, find $\frac{d^2y}{dx^2}$.

Solution: Use the Product Rule to find the first derivative:

$$\frac{dy}{dx} = (1) \sin(x) + x \cos(x) = \sin(x) + x \cos(x)$$

Take the derivative of this function to get the second derivative:

$$\frac{d^2y}{dx^2} = \cos(x) + (1) \cos(x) + x (-\sin(x)) = 2 \cos(x) - x \sin(x)$$

YOU'VE GOT PROBLEMS

Problem 3: Given k(x) = $\sin^2(3x)$, find k"(x).

Example 8: Given $p(x) = e^{\cos(x)}$, find $p'''(x)$, the third derivative of $p(x)$.

Solution: The first derivative is an application of the Chain Rule:

$$p'(x) = -e^{\cos(x)}\sin(x)$$

The second derivative requires the Product Rule (I use brackets [ ] to indicate the derivative so it is easier to follow):

$$p''(x) = [e^{\cos(x)}\sin(x) + (-e^{\cos(x)})[\cos(x) = e^{\cos(x)}(\sin^2(x) - \cos(x))$$

The third derivative also involves the Product Rule:

$$p'''(x) = e^{\cos(x)}\left(-\sin^3(x) + 3\sin(x)\cos(x) + \sin(x)\right)$$

Example 9: Find the equation of the line tangent to $q(t) = \sin^2(2t)$ at $t = \frac{\pi}{12}$.

Solution: We need two pieces of information to write the equation of the tangent line:

- The point through which the line passes
- The slope of the line

We find the point through which the line passes by evaluation of $q(t)$ at $t = \frac{\pi}{12}$,

$q\left(\frac{\pi}{12}\right) = \sin^2\left(2\left(\frac{\pi}{12}\right)\right) = \sin^2\left(\frac{\pi}{6}\right) = \frac{1}{4}$. The derivative of $q(t)$ is $q'(t) = 4\sin(2t)\cos(2t) = 2\sin(4t)$ so the

slope of the tangent line is $q'\left(\frac{\pi}{12}\right) = 4\sin\left(4\left(\frac{\pi}{12}\right)\right) = 4\sin\left(\frac{\pi}{3}\right) = \sqrt{3}$. Using the point-slope form for

the line, the equation of the tangent line is $y - \frac{1}{4} = \sqrt{3}\left(x - \frac{\pi}{12}\right)$.

YOU'VE GOT PROBLEMS

Problem 4: Write the equation for the line tangent to $w(z) = \frac{z^2+1}{3z-5}$ at $z = 2$.

Implicit Differentiation

Although the equation for a circle is not a function, we should be able to write the equation for a line tangent to the circle at the point of our choosing. To accomplish this, we could simply identify which quadrant of the circle is sought and then limit the domain and range of the equation $y = \sqrt{r^2 - x^2}$ to manufacture the function needed. However, there are any number of nonfunctional relations that we should also be able to examine without having to perform (sometimes difficult) algebraic manipulations in order to achieve this goal. To that end, we apply implicit differentiation.

> ✏️ **CRITICAL POINT**
>
> The notation $y = f'(x)$ and $\frac{dy}{dx}$ indicates that y is dependent on x.

Example 10: Given the equation $x^2 - 4xy - 3y^2 = 21$, find the equation of the line tangent to the curve at the point $(3, -2)$.

Solution: We know the point through which the line passes; now we need to find the slope of the line by first finding the derivative, $\frac{dy}{dx}$.

This is where we can talk about the beauty of the notation $\frac{dy}{dx}$ where we are told to find the derivative of y with respect to x. That is, we are told that y depends on x.

Thus, when we take the derivative of $x^2 - 4y - 3xy^2 = 21$ we are told that x is the independent variable and that y depends on x. Whenever we differentiate a term in y, we need to apply the Chain Rule. The derivative of the equation is $2x - \left(4y + 4x\frac{dy}{dx}\right) - 6y\frac{dy}{dx} = 0$. Solve for $\frac{dy}{dx}$ to get

$\frac{dy}{dx} = \frac{2x - 4y}{6y + 4x}$. At the point $(3, -2)$, the slope of the tangent line is $\frac{dy}{dx} = \frac{2(3) - 4(-2)}{6(-2) + 4(3)} = \frac{14}{0}$, which is undefined, indicating that the tangent line is vertical. The equation of the tangent line is $x = 3$.

Example 11: Find $\frac{dy}{dx}$ given $\sin(xy) + \cos(y) = x$.

Solution: Use the Chain Rule and Product Rule on the first term to get $\cos(xy)\left[y + x\frac{dy}{dx}\right] - \sin(y)\frac{dy}{dx} = 1$. Gather all the terms in $\frac{dy}{dx}$:

$y\cos(xy) + x\cos(xy)\frac{dy}{dx} - \sin(y)\frac{dy}{dx} = 1$ becomes $\left(x\cos(xy) - \sin(y)\right)\frac{dy}{dx} = 1 - y\cos(xy)$. Divide

to find $\frac{dy}{dx} = \frac{1 - y\cos(xy)}{x\cos(xy) - \sin(y)}$.

Example 12: Find $\frac{d^2y}{dx^2}$ if $e^y + \sin(y) = x^2$.

Solution:

1. First find the value of $\frac{dy}{dx}$, $e^y \frac{dy}{dx} + \cos(y)\frac{dy}{dx} = 2x$. Therefore, $\left(e^y + \cos(y)\right)\frac{dy}{dx} = 2x$ and

   $\frac{dy}{dx} = \frac{2x}{e^y + \cos(y)}$.

2. Take the derivative of this expression to find $\frac{d^2y}{dx^2}$.

3. Use the Quotient Rule to find $\frac{d^2y}{dx^2} = \frac{2\left(e^y + \cos(y)\right) - 2x\left[e^y\frac{dy}{dx} - \sin(y)\frac{dy}{dx}\right]}{\left(e^y + \cos(y)\right)^2}$.

4. Factor the $\frac{dy}{dx}$ from the right-hand term in the numerator to get

   $\frac{d^2y}{dx^2} = \frac{2\left(e^y + \cos(y)\right) - 2x\left[e^y - \sin(y)\right]\frac{dy}{dx}}{\left(e^y + \cos(y)\right)^2}$.

5. Use the result of the first derivative to get $\frac{d^2y}{dx^2} = \frac{2\left(e^y + \cos(y)\right) - 2x\left[e^y - \sin(y)\right]\left(\frac{2x}{e^y + \cos(y)}\right)}{\left(e^y + \cos(y)\right)^2}$.

6. Multiply numerator and denominator by the term $e^y + \cos(y)$ to get

   $\frac{d^2y}{dx^2} = \frac{2\left(e^y + \cos(y)\right)^2 - 4x^2\left[e^y - \sin(y)\right]}{\left(e^y + \cos(y)\right)^3}$.

YOU'VE GOT PROBLEMS

Problem 5: Find $\frac{d^2y}{dx^2}$ if $3x^2 - xy - 4y^2 = 8$.

Mean Value Theorem

Because the derivative is based on a limit definition, at times the limit might not exist. This is certainly true when the function has a point of discontinuity. It is also true whenever the graph of the function contains a cusp, as does the graph of $f(x) = x^{\frac{2}{3}}$ at $x = 0$. The crux of the issue is that a continuous function need not be differentiable at all points. However, a function that is differentiable at all points in an interval is continuous with the interval.

The most important theorem in calculus is the Fundamental Theorem of Calculus (which we will review in Chapter 3). The second most important theorem is the *Mean Value Theorem*.

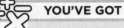

DEFINITION

The **Mean Value Theorem** states that if a function f(x) is defined and continuous on [a,b] and differentiable of (a, b) there is a value of c in the interval (a,b) so that the instantaneous rate of change at c is equal to the average rate of change over [a,b]. That is, there exists a value of c in (a,b) so that $f'(c) = \frac{f(b) - f(a)}{b - a}$.

You can see that the Mean Value Theorem does not apply to the function $f(x) = x^{\frac{2}{3}}$ on [–1, 1]. f(–1) = f(1) = 1 so the average rate of change for f(x) on this interval is 0. However, the derivative of f(x) is $\frac{2}{3}x^{-\frac{1}{3}}$ and this expression can never equal 0. Remember that the graph of f(x) has a cusp at x = 0 and, therefore, the function does not satisfy the condition that the function be differentiable in the interval (–1, 1).

YOU'VE GOT PROBLEMS

Problem 6: Find a value of c on the interval [1,5] that satisfies the Mean Value Theorem for the function f(x) = x^3 + 3.

Relative Extremes and Concavity

As a measure of the instantaneous rate of change, the derivative indicates whether a function is:

- Increasing (The derivative is positive.)

- Decreasing (The derivative is negative.)

- Stationary (The derivative is 0.)

Example 13: Suppose f(x) is differentiable over its domain with f'(a) = f'(b) = f'(c) = 0, f'(x) > 0 when x < a, b < x < c, and x > c, as shown in the following figure.

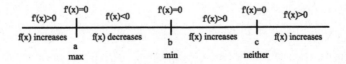

Figure 2.3

The signs analysis for the first derivative is to determine when a function is increasing (the derivative is positive) and when it is decreasing (the derivative is negative).

Solution: The *sign analysis* shows that the graph of f(x) has a relative maximum at $x = a$ because ...

1. The graph increases until it reaches the point $(a, f(a))$ and then decreases.

2. The graph continues to decrease until it reaches the point $(b, f(b))$ and then it rises, making the f(b) a relative minimum.

3. The graph continues to rise until $x = c$ when it momentarily levels off then continues to rise.

4. In this case, there is neither a relative minimum nor maximum at $x = c$.

5. This analysis is referred to as the *First Derivative Test*.

> **DEFINITION**
>
> A **signs analysis** uses a number line to show the intervals in which the derivative of the function is positive or negative as well as the points at which the derivative is equal to 0 or fails to exist. The **First Derivative Test** is used to test critical points, points at which the value of the derivative is equal to 0, or when the derivative fails to exist at a specific point even though the function is defined at that point. Relative extremes occur when the sign of the derivative changes from one side of the critical point to the other.

The curvature of a graph is based on how fast the rate of change is changing. That is, the curvature is based on the derivative of the first derivative, called the second derivative. (The second derivative is designated with a double prime in Newton's notation, f"(x), and with a squared notation in Leibniz's notation, $\frac{d\left(\frac{dy}{dx}\right)}{dx} = \frac{d^2y}{dx^2}$. Don't get hung up on the algebraic equivalence of that fraction, this is not an algebraic manipulation but notation.)

The graph is said to be concave up (is capable of "holding water") when f"(x) > 0 (like a section of a parabola opening up) and is concave down ("spilling water") when f"(x) < 0 (like a section of a downward opening parabola). The point at which the graph changes concavity is called the *point of inflection*.

> **BE AWARE**
>
> The graph of f(x) has a relative maximum as $x = a$ but the maximum value is f(a). One phrase identifies when the extreme occurs while the other gives the extreme value.

Example 14: Given $f(x) = x^5 - 10x^4 - 15x^3 + 2$, find all points for which $f'(x) = 0$ and for which $f''(x) = 0$.

Solution: The first derivative, $f'(x) = 5x^4 - 40x^3 - 45x^2 = 5x^2 (x^2 - 8x - 9)$. Consequently, $f'(x) = 0$ when $x = -1$, 0, or 9. The second derivative, $f''(x) = 20x^3 - 120x^2 - 90x = 10x (2x^2 - 12x - 9)$. The second derivative equals 0 when $x = -0.67$, 0, or 6.67. The graph of $f(x)$ has a point of inflection at $x = 0$ (and a horizontal tangent line as well at $x = 0$).

A sign analysis for $f'(x)$ shows that $f(x)$ has a relative maximum at $x = -1$ and a relative minimum at $x = 9$.

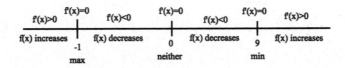

Figure 2.4

The signs analysis for the first derivative of $f(x) = x^5 - 10x^4 - 15x^3 + 2$.

A sign analysis for $f''(x)$ shows that $f(x)$ is concave down when $x < -0.67$ and $0 < x < 6.67$ and that $f(x)$ is concave up when $-0.67 < x < 0$ and $x > 6.67$.

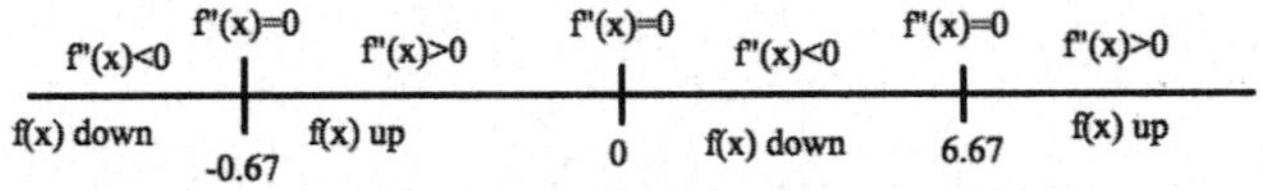

Figure 2.5

The signs analysis for the second derivative of $f(x) = x^5 - 10 x^4 - 15 x^3 + 2$. *The graph of* $y = f(x)$ *is concave up when* $f''(x) > 0$ *and concave down when* $f''(x) < 0$.

Notice that the graph is concave down when $f(x)$ has a maximum and concave up when the graph has a minimum. This is the essence of the *Second Derivative Test.* If $f'(c) = 0$ and $f''(c) < 0$, then $f(x)$ has a relative maximum at $x = c$. If $f'(c) = 0$ and $f''(x) > 0$, then $f(x)$ has a relative minimum at $x = c$. If both $f'(c) = 0$ and $f''(c) = 0$, the test fails.

YOU'VE GOT PROBLEMS

Problem 7: The graph of the derivative of f(x) is shown. Determine when the function has its relative extreme values and the nature of those extremes.

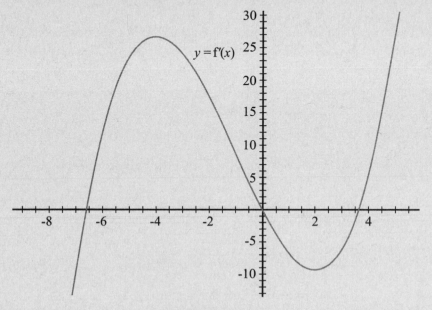

$y = f'(x)$

Figure 2.6
This is the sketch of f'(x), *not* f(x).

Applications of the Derivative

There are times when evaluating a limit leads to an indeterminate result such as $\frac{0}{0}$ or $\pm\frac{\infty}{\infty}$. In the case of $\frac{0}{0}$, the factorization is either tedious or not possible. In many cases, you can use derivatives to evaluate the limits.

DEFINITION

Given $\lim\limits_{x \to c} \frac{f(x)}{g(x)}$, with both f(x) and g(x) differentiable at $x = c$. If $\lim\limits_{x \to c} \frac{f(x)}{g(x)} = \frac{0}{0}$ or $\pm\frac{\infty}{\infty}$, then $\lim\limits_{x \to c} \frac{f(x)}{g(x)} = \lim\limits_{x \to c} \frac{f(x)}{g(x)}$. This is called *L'Hopital's Rule*.

Example 15: Evaluate $\lim\limits_{x \to 0} \frac{\sin(x)}{\tan(x)}$.

Solution: Because $\lim\limits_{x \to 0} \frac{\sin(x)}{\tan(x)} = \frac{0}{0}$, L'Hopital's Rule applies. The derivative of $\sin(x)$ is $\cos(x)$ and the derivative of $\tan(x)$ is $\sec^2(x)$ so $\lim\limits_{x \to 0} \frac{\sin(x)}{\tan(x)} = \lim\limits_{x \to 0} \frac{\cos(x)}{\sec^2(x)} = \frac{1}{1} = 1$.

You could have used trig identities to solve Example 15. Using $\tan(x)$ as the quotient of $\sin(x)$ and $\cos(x)$, the original fraction reduces to $\cos(x)$ and is easily solved. The next problem does not have an "easier" approach.

Example 16: Evaluate $\lim\limits_{x \to 0} \frac{e^x - 1}{x^2}$.

Solution: Because $\lim\limits_{x \to 0} \frac{e^x - 1}{x^2} = \frac{0}{0}$, L'Hopital's Rule applies. The derivative of $e^x - 1$ is e^x and the derivative of x^2 is $2x$ so $\lim\limits_{x \to 0} \frac{e^x - 1}{x^2} = \lim\limits_{x \to 0} \frac{e^x}{2x} = \frac{1}{0}$, and the limit fails to exist.

YOU'VE GOT PROBLEMS

Problem 8: Evaluate $\lim\limits_{x \to 0} \frac{1 - \cos(x)}{\sin(x)}$.

Derivatives are used to determine optimal solutions to applied problems (greatest profit, minimum cost, least area, largest volume to name a few). The issue to solving many of these problems is in establishing the correct equation.

Example 17: A company has determined that the daily revenue it can earn from selling n units of its product is $-0.05n^2 + 200n$ while the daily cost of producing these n units is $125n + 1100$. How many units must be produced and sold to maximize profit?

Solution: The daily profit function is the difference between revenue and cost so $P(n) = -0.05n^2 + 200n - (125n + 1100) = -0.05n^2 + 75n - 1100$. $P'(n) = -0.1n + 75$. Set $P'(n) = 0$ and solve to get $n = 750$. The company should produce and sell 750 units daily.

Example 18: A manufacturer has its factory at point A and its distribution center at point B, on opposite banks of a river that is 1 mile wide. Point B is also 20 miles downstream from point A. The company is planning to build a dock for a ferry that will allow trucks to transport goods from the factory to the distribution center. If the ferry can travel at 5 mph across the river and the trucks can travel on the road along side the river at 30 mph, where should the dock be built to minimize the time to transport the goods from point A to point B?

Solution: A diagram of the situation is helpful:

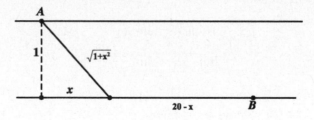

Figure 2.7
A diagram representing the route that the truck will take to get from the factory to the distribution point.

1. The ferry can land anywhere between the point directly across the shore from point A to point B.

2. Label the distance from the point across from A to the ferry dock as x.

3. The distance the ferry will travel is given by the Pythagorean Theorem and the distance the truck will travel is $20 - x$.

4. Ignoring the time it takes the truck to get off the ferry (because that will be the same no matter where the dock is built), the total travel time is $t(x) = \frac{\sqrt{1+x^2}}{5} + \frac{20-x}{30}$.

5. Take the derivative of this function to get $t'(x) = \frac{x}{5\sqrt{1+x^2}} - \frac{1}{30}$.

6. Set t'(x) equal to 0 to get $30x = 5\sqrt{1+x^2}$.

7. Divide by 5, $6x = \sqrt{1+x^2}$.

8. Square both sides of the equation, and combine terms to get $35x^2 = 1$ and $x = \frac{1}{\sqrt{35}} = 0.169$.

9. Construct the dock 0.169 miles downstream from point A to minimize transportation time.

YOU'VE GOT PROBLEMS

Problem 9: A rectangle *ABCD* with sides parallel to the coordinate axes is inscribed in the region enclosed by the graph of $y = 4 - x^2$ and the coordinate axes, as shown in the figure. Find the coordinates of point *A* so the area of the rectangle is a maximum.

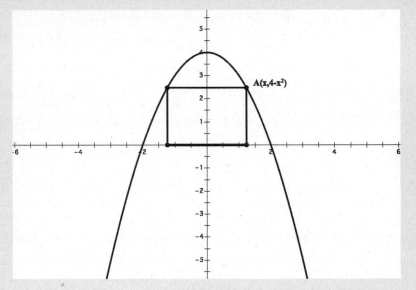

Figure 2.8

A rectangle is inscribed inside the graph of $y = 4 - x^2$ and uses the x-axis as a side of the rectangle.

Related Rates

As discussed in the "Parametric Equations" section of Chapter 1, we sometimes need to think of the terms in our problems as functions of time. For instance, think about a 25-foot-long ladder leaning against the wall. If you measure how far the bottom of the ladder is from the base of the wall, you can use the Pythagorean Theorem to determine how high the ladder is above the ground. What happens to the top of the ladder as the bottom of the ladder slides away from the wall? Does it slide at the same rate? In this case, the positions of the bottom and top of the ladder are dependent upon the amount of time that has passed.

Example 19: Suppose the base of the ladder is sliding away from the wall at 1 foot per second. At what rate is the top of the ladder falling when the base of the ladder is 7 feet from the wall? 15 feet from the wall?

Solution:

1. If you call x the distance from the base of the wall and y the height of the ladder above the ground, the Pythagorean relationship gives $x^2 + y^2 = 625$.

2. The base of the ladder is sliding away from the wall at 1 foot per second, that is, $\frac{dy}{dx} = 1$.

3. Differentiate the equation to get $2x\frac{dx}{dt} + 2y\frac{dy}{dt} = 0$ or $\frac{dy}{dt} = -\frac{x}{y}\frac{dx}{dt}$.

4. When the base of the ladder is 7 feet from the wall, the top of the ladder is 24 feet above the ground so the top of the ladder is falling at a rate of $\frac{-7}{24}$ feet per second but when the base of the ladder is 15 feet from the wall, the rate of descent of the top of the ladder is $\frac{-15}{20} = \frac{-3}{4}$ feet per second.

Example 20: A right circular cone and a hemisphere have the same base, and the cone is inscribed in the hemisphere. The figure is expanding in such a way that the combined surface area of the hemisphere and its base is increasing at a constant rate of 18 square inches per second. At what rate is the volume of the cone changing at the instant when the radius of the common base is 4 inches? (Area of a Sphere: $4\pi r^2$ and the volume of a cone is $\frac{1}{3}\pi r^2 h$.)

Solution:

1. If the cone is inscribed within the hemisphere, then the height of the cone must be equal to the radius of the sphere, meaning the volume of the cone is $V = \frac{1}{3}\pi r^3$.

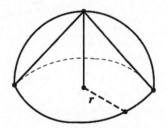

Figure 2.9
A cone is inscribed within a hemisphere. The bases of the two figures are the same.

2. The combined area of the hemisphere and its base is $3\pi r^2$ (half the area of the sphere plus the area of the circle with radius r).

3. This area is increasing at the rate of 18 square inches per second. That is, $\frac{dA}{dt} = 18$.

4. We are asked to find the rate of change of the volume, $\frac{dV}{dt}$.

5. $A = 3\pi r^2$ yields $\frac{dA}{dt} = 6\pi r \frac{dr}{dt}$.

6. When $r = 4$, $18 = 24\pi \frac{dr}{dt}$, so $\int_0^{\sqrt{t}} \sin\left(\cos\left(t'\right)\right) dt$.

7. $\frac{dV}{dt} = \pi r^2 \frac{dr}{dt}$, so $\frac{dV}{dt} = \left(16\pi\right)\left(\frac{3}{4\pi}\right) = 12$ cubic inches per second.

YOU'VE GOT PROBLEMS

Problem 10: A ladder 15 feet long is leaning against a building so end x is on level ground and end y is on the wall as shown in Figure 2.10. x is moved away from the building at the constant rate of $\frac{1}{2}$ foot per second.

(a) Find the rate at which the length OY is changing when x is 9 feet from the building.

(b) Find the rate of change of the area of $\triangle XOY$ when x is 9 feet from the building.

Figure 2.10
A diagram showing a ladder leaning against a building. How fast does point y slide down the wall as point x moves away from the base?

The Least You Need to Know

- The differentiation rules enable you to calculate the derivative of a function.

- The Product, Quotient, and Chain Rules help you differentiate functions.

- The notation $\frac{dy}{dx}$ indicates that x is the independent variable and y is the dependent variable.

- The first derivative test is used to determine the location of maxima and minima by determining when the function is increasing and decreasing.

- When the variables are functions of time, use related rates to determine the derivatives.

Definite and Indefinite Integrals

In elementary school, you learned that the opposite of addition is subtraction. You also learned that the opposite of multiplication is division.

In high school algebra, you learned that the opposite of squaring a number was the square root of the number. You then learned that the opposite process of a logarithm is exponentiation (sometimes referred to as an antilog).

In Calculus I, you learned how to take the derivatives of a variety of functions. The question then becomes how do you find the function that you started with if you know the derivative of that function?

In This Chapter

* Working with indefinite integrals—don't forget the $+ C$
* Making integrals easier to see with u-substitutions
* Understanding the Fundamental Theorem of Calculus
* Applying plane geometry with numerical approximations

Indefinite Integrals

Suppose you know that the derivative of the function $y = f(x)$ is $f'(x) = 2x$. What is the original function? The correct answer is that without more information, you don't know. Yes, it is true that the derivative of $y = x^2$ is $y = 2x$, but so is the derivative of $y = x^2 + 3$, $y = x^2 + 300$, and $y = x^2 - \sqrt{5}$. There are an infinite number of functions of the variety $y = x^2 + C$ for which the derivative will be $2x$. The process of finding the antiderivative is also known as *integration*.

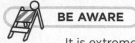

BE AWARE

It is extremely important for you to remember that the antiderivative of a function can be any of a host of functions. You also should remember to add the constant of integration to your answers.

Example 1: Given $f'(x) = 4x + 5$, find an expression for $f(x)$.

Solution: The derivative for x^2 is $2x$, so it stands to reason that the antiderivative of $4x$ is $2x^2$. The antiderivative of the constant 5 is $5x$. Therefore, $f(x) = 2x^2 + 5x + C$.

Example 2: Given $g'(x) = 5x^3 + 6x^2 + e$, determine an expression for $g(x)$.

Solution: The process for dealing with this problem is the same as in the previous example. The part that gets a little sloppy is the coefficients.

The derivative of x^4 is $4x^3$ so the antiderivative for $5x^3$ must be $\frac{5}{4}x^4$. The antiderivative for $6x^2$ is $2x^3$. (Take the derivative of $2x^3$ to verify this.) The last term in $g'(x)$ is the constant e and the antiderivative is the linear expression ex. Therefore, $g(x) = \frac{5}{4}x^4 + 2x^3 + ex + C$.

YOU'VE GOT PROBLEMS

Problem 1: What is the antiderivative of $k'(x) = 8x^3 + 5x - \frac{1}{\sqrt{x+2}}$?

You should be familiar with a number of differentiation formulas from Calculus I. These are listed in the left column of the following table with the corresponding antiderivatives in the right column.

Derivative	Antiderivative
$y = x^n \rightarrow \frac{dy}{dx} = nx^{n-1}$	$\frac{dy}{dx} = x^n \rightarrow y = \frac{1}{n+1} x^n + 1 + C \ (n \neq -1)$
$y = e^x \rightarrow \frac{dy}{dx} = e^x$	$\frac{dy}{dx} = e^x \rightarrow y = e^x + C$
$y = \ln(x) \rightarrow \frac{dy}{dx} = \frac{1}{x}$	$\frac{dy}{dx} = \frac{1}{x} \rightarrow y = \ln|x| + C$
$y = b^x \rightarrow \frac{dy}{dx} = \ln(b) \times b^x$	$\frac{dy}{dx} = b^x \rightarrow y = \frac{1}{\ln(b)} b^x + C$
$y = \log_b(x) \rightarrow \frac{dy}{dx} = \frac{1}{x\ln(b)}$	$\frac{dy}{dx} = \frac{1}{x\ln(b)} \rightarrow y = \frac{\ln|x|}{\ln(b)} + C = \log_b |x| + C$
$y = \sin(x) \rightarrow \frac{dy}{dx} = \cos(x)$	$\frac{dy}{dx} = \cos(x) \rightarrow y = \sin(x) + C$
$y = \cos(x) \rightarrow \frac{dy}{dx} = -\sin(x)$	$\frac{dy}{dx} = \sin(x) \rightarrow y = -\cos(x) + C$
$y = \tan(x) \rightarrow \frac{dy}{dx} = \sec^2(x)$	$\frac{dy}{dx} = \sec^2(x) \rightarrow y = \tan(x) + C$
$y = \sec(x) \rightarrow \frac{dy}{dx} = \sec(x)\tan(x)$	$\frac{dy}{dx} = \sec(x)\tan(x) \rightarrow y = \sec(x) + C$
$y = \csc(x) \rightarrow \frac{dy}{dx} = -\csc(x)\cot(x)$	$\frac{dy}{dx} = \csc(x)\cot(x) \rightarrow y = -\csc(x) + C$
$y = \cot(x) \rightarrow \frac{dy}{dx} = -\csc^2(x)$	$\frac{dy}{dx} = \csc^2(x) \rightarrow y = -\cot(x) + C$
$y = \sin^{-1}(x) \rightarrow \frac{dy}{dx} = \frac{1}{\sqrt{1-x^2}}$	$\frac{dy}{dx} = \frac{1}{\sqrt{1-x^2}} \rightarrow y = \sin^{-1}(x) + C$
$y = \cos^{-1}(x) \rightarrow \frac{dy}{dx} = -\frac{1}{\sqrt{1-x^2}}$	$\frac{dy}{dx} = -\frac{1}{\sqrt{1-x^2}} \rightarrow y = \cos^{-1}(x) + C$
$y = \tan^{-1}(x) \rightarrow \frac{dy}{dx} = \frac{1}{1+x^2}$	$\frac{dy}{dx} = \frac{1}{1+x^2} \rightarrow y = \tan^{-1}(x) + C$

Example 3: Find the antiderivative for $f'(x) = \sqrt{4x+3}$.

Solution:

1. Rewrite the radical as an exponential expression, $f'(x) = (4x+3)^{1/2}$.

2. Using rule 1, the antiderivative should be of the form $y = \frac{1}{1+\frac{1}{2}} x^{\frac{1}{2}+1} = \frac{2}{3} x^{\frac{3}{2}}$.

3. The derivative of $y = \frac{2}{3}(4x+3)^{\frac{3}{2}}$ is $\left(\frac{2}{3}\right)\left(\frac{3}{2}(4x+3)^{\frac{1}{2}} \times 4\right) = 6(4x+3)^{\frac{1}{2}}$.

4. Since this is 6 times the value of the initial expression, it stands to reason that
$$f(x) = \frac{1}{6}(4x+3)^{\frac{3}{2}} + C.$$

Example 4: If $\frac{dy}{dx} = 5\sin(x) + 4\sec^2(x) + e^{2x}$, find an expression for y.

Solution:

1. The antiderivative for $\sin(x)$ is $-\cos(x)$, the antiderivative for $\sec^2(x)$ is $\tan(x)$, and the antiderivative for e^x is e^x.

2. The derivative for e^{2x} is $2e^{2x}$ so the antiderivative for e^{2x} is $\frac{1}{2}e^{2x}$.

3. Therefore, $y = -5\cos(x) + 4\tan(x) + \frac{1}{2}e^{2x} + C$.

Example 5: If $f'(x) = \frac{1}{x-4} + \sqrt{x+4}$ and $f(5) = 2$, find the expression for $f(x)$.

Solution:

1. The antiderivative of $\frac{1}{x-4}$ is $\ln|x{-}4|$.

2. The antiderivative for $\sqrt{x+4}$ is $\frac{2}{3}(x+4)^{\frac{3}{2}}$.

3. Therefore, $f(x) = \ln|x-4| + \frac{2}{3}(x+4)^{\frac{3}{2}} + C$.

4. You can now use the information that $f(5) = 2$ to get $2 = \ln|5-4| + \frac{2}{3}(5+4)^{\frac{3}{2}} + C$.

5. Solve for C: $2 = \ln|1| + \frac{2}{3}(9)^{\frac{3}{2}} + C$, which becomes $2 = 18 + C$ and $C = -16$.

6. Therefore, $f(x) = \ln|x| + \frac{2}{3}(x+4)^{\frac{3}{2}} - 16$.

CRITICAL POINT

Mathematics is famous (or infamous, depending upon your opinion) for using notation to represent a number of words. Another example of the simplification of language is the use of the symbol, $\int$, a Gothic S, to represent the antiderivative. Rather than write, "If

$$\int 2\cos(x) + \sqrt{3}\sec(x)\tan(x) - \frac{6}{6x+1-\pi}\,dx \quad \text{find an expression for } y,\text{" it}$$

is briefer to write $\int 2\cos(x) + \sqrt{3}\sec(x)\tan(x) - \frac{6}{6x+1-\pi}\,dx$ to say the same thing. The differential dx at the end of the phrase indicates which symbol is the independent variable of the problem. While the command is to find an antiderivative, the notation is read "Find the integral of

$$2\cos(x) + \sqrt{3}\sec(x)\tan(x) - \frac{6}{6x+1-\pi} \quad \text{with respect to } x.\text{"}$$

Let's try another example to find a specific function given its derivative and a point on the function.

Example 6: Given $f(x) = \int 2\cos(x) + \sqrt{3}\sec(x)\tan(x) - \frac{6}{6x+1-\pi}\,dx$ and $f\left(\frac{\pi}{6}\right) = 9$ find f(x).

Solution:

1. The antiderivative of $2\cos(x)$ is $2\sin(x)$, of $\sqrt{3}\sec(x)\tan(x)$ is $\sqrt{3}\sec(x)$, and of $\frac{6}{6x+1-\pi}$ is $\ln|6x + 1 - \pi|$.

2. Consequently, $\int 2\cos(x) + \sqrt{3}\sec(x)\tan(x) - \frac{6}{6x+1-\pi}\,dx = 2\sin(x) + \sqrt{3}\sec(x) - \ln|6x + 1 - \pi| + C$.

3. Given $f\left(\frac{\pi}{6}\right) = 9$, we get:

$$9 = 2\sin\left(\tfrac{\pi}{6}\right) + \sqrt{3}\sec\left(\tfrac{\pi}{6}\right) - \ln\left|6\left(\tfrac{\pi}{6}\right) + 1 - \pi\right| + C$$

$$9 = 2\left(\tfrac{1}{2}\right) + \sqrt{3}\left(\tfrac{2}{\sqrt{3}}\right) - \ln|\pi + 1 - \pi| + C$$

$$9 = 1 + 2 - 0 + C$$

$$C = 6$$

4. Therefore, $f(x) = 2\sin(x) + \sqrt{3}\sec(x) - \ln|6x + 1 - \pi| + 6$.

Example 7: Evaluate $\int \sin^2(x)\,dx$.

Solution:

1. The integrand $\sin^2(x)$ is not in our list of derivatives/antiderivatives that we need to know.

2. What we do know about $\sin^2(x)$ is that it is part of one of the formulas for $\cos(2x)$, $\cos(2x) = 1 - 2\sin^2(x)$.

3. Solve this equation for $\sin^2(x)$ to get $\sin^2(x) = \frac{1}{2} - \frac{1}{2}\cos(2x)$.

4. Change the integrand to this equivalent value $- \int \sin^2(x)\,dx = \int \frac{1}{2} - \frac{1}{2}\cos(2x)dx =$
$\frac{1}{2}x - \frac{1}{2}\left(\frac{1}{2}\sin(2x)\right) + C = \frac{1}{2}x - \frac{1}{4}\sin(2x) + C$.

u-Substitutions

There are times when the expression within the integral, called the *integrand*, is "involved." That is, the integral fits one of the basic patterns. However, due to the effect of the Chain Rule, there are seemingly more terms to handle. The process of u-substitution restores the integrand to a form in which the antiderivative is a known formula.

> **DEFINITION**
>
> The **integrand** is an expression in which the antiderivative is being found in an integral. That is, f(x) is the integrand in the integral $\int f(x)\,dx$.

Example 8: Evaluate $\int (8x + 12)(x^2 + 3x + 1)^4\,dx$.

Solution:

1. The integrand has the pattern of a power rule ($y = x^n$) problem.

2. Transform the problem by letting $u = x^2 + 3x + 1$.

3. Differentiate this to get $\frac{dy}{dx} = 2x + 3$, which becomes $du = (2x + 3)dx$.

4. Notice that $8x + 12$ is 4 times the value of $2x + 3$ so $4du = (8x + 12)dx$ and the original problem $\int (8x + 12)(x^2 + 3x + 1)^4\,dx$ now becomes $\int 4u^4\,du$ or $4\int u^4\,du$.

5. The antiderivative of u^4 is $\frac{1}{5}u^5$.

6. Transform back to the original problem to get $\int (8x + 12)(x^2 + 3x + 1)^4\,dx =$
$\frac{4}{5}(x^2 + 3x + 15)^5 + C$.

Example 9: Evaluate $\int \frac{2x+4}{x^2+4x+3}\,dx$.

Solution:

1. The denominator contains a polynomial that is not being raised to a power, so it does not fit the x^n rule.

2. The numerator is the derivative of the denominator. Just as you did in Algebra I, you define your variable.

3. Let $u = x^2 + 4x + 3$. Take the derivative of u with respect to x to get $\frac{dy}{dx} = 2x + 4$ or $du = (2x + 4)dx$.

4. Make the changes in the integral. $\int \frac{2x+4}{x^2+4x+3}\,dx$ becomes $\int \frac{1}{u}\,du$, which equals $\ln|u|$.

5. We then get the answer to the problem, $\int \frac{2x+4}{x^2+4x+3}\,dx = \ln|x^2 + 4x + 3| + C$.

Example 10: Evaluate $\int x^3 \cos^5\!\left(x^4\right)\sin\!\left(x^4\right)dx$.

Solution:

1. The derivative of x^4 is $4x^3$, and we see x^3 in the problem.

2. We also know that the derivative of the cosine function is the negative of the sine function and that they are both in the problem.

3. Lastly, the cosine function is being raised to a power. It is almost always the case that the most "complicated" term (the term composed with the most functions) will be the place in which the transformation will occur.

4. The term $\cos^5(x^4)$ consists of three functions:

   - Raising a term to the 4th power

   - Taking the cosine of that result

   - Raising this last result to the 5th power

5. The transformation for u uses all but the last of these functions. That is, let $u = \cos(x^4)$. Rather than writing the fraction $\frac{du}{dx}$ each time and then multiplying by dx, let's just jump to the $du =$ step in the process. $du = -4x^3\sin(x^4)dx$ so $\frac{-1}{4}\,du = x^3\sin(x^4)dx$. The original problem becomes $\frac{-1}{4}\int u^5\,du = \frac{-1}{24}u^6$.

6. Therefore, $\int x^3 \cos^5\!\left(x^4\right)\sin\!\left(x^4\right)dx = \frac{-1}{24}\cos^6(x^4) + C$.

Problems get to be a little more interesting when not all of the integrand can be easily converted using u-substitution. For example, evaluate $\int x \sec^5\left(x^2\right)\tan\left(x^2\right)dx$.

1. We can account for the leading x because it is the derivative of x^2.

2. If we let $u = \sec(x^2)$, we run into the problem that $du = 2x\sec(x^2)\tan(x^2)\,dx$. Fortunately, this is an easy problem to overcome because the expression for du tells us that we need to think of the problem as $\int x \sec^5\left(x^2\right)\tan\left(x^2\right)dx$ and transform this to $\frac{1}{2}\int u^4\,du$.

3. We can then determine that $\int x \sec^5\left(x^2\right)\tan\left(x^2\right)dx = \frac{1}{10}\sec^5(x^2) + C$.

Consider the problem $\int \frac{1-x}{\sqrt{1-x^2}}dx$. It would seem reasonable to let $u = 1 - x^2$ so that $du = -2x\,dx$. Unfortunately, that does not account for the 1 in the numerator. This is where you must recall all the basic anti-differentiation formulas.

1. Because $\int \frac{1}{\sqrt{1-x^2}}dx = \sin^{-1}(x)$, you can rewrite the original problem as $\int \frac{1}{\sqrt{1-x^2}} - \frac{x}{\sqrt{1-x^2}}dx$ and then as $\int \frac{1}{\sqrt{1-x^2}}dx - \int \frac{x}{\sqrt{1-x^2}}dx$.

2. The first integral is the inverse sine and the second is $\sqrt{1-x^2}$ so $\int \frac{1}{\sqrt{1-x^2}}dx = \sin^{-1}(x) + \sqrt{1-x^2} + C$.

YOU'VE GOT PROBLEMS

Problem 2: Evaluate $\int x^2 \cos\left(x^3\right)e^{\sin\left(x^3\right)}dx$.

Here's a problem that looks unlike the other "complicated" problems that you just did but is interesting nonetheless.

Example 11: Evaluate $\int \tan(x)\,dx$.

Solution: There are two ways to approach this problem:

1. First, because $\tan(x) = \frac{\sin(x)}{\cos(x)}$, rewrite the integral as $\int \frac{\sin(x)}{\cos(x)}dx$.

2. Let $u = \cos(x)$ so that $du = -\sin(x)\,dx$.

3. The problem now becomes $\int \frac{-1}{u}du$, which equals $-\ln|u| + C$ so $\int \tan(x)\,dx = -\ln|\cos(x)| + C$.

The second solution is to rewrite $\int \tan(x)\,dx$ as $\int \frac{\sec(x)\tan(x)}{\sec(x)}\,dx$ by multiplying numerator and denominator by $\sec(x)$.

1. Let $u = \sec(x)$ so that $du = \sec(x)\tan(x)dx$.

2. The problem now becomes $\int \frac{1}{u}\,du$, which equals $\ln|u| + C$ so that $\int \tan(x)\,dx = \ln|\sec(x)| + C$.

CRITICAL POINT

Add $\int \tan(x)\,dx = \ln|\sec(x)| + C$ to your list of important antiderivatives you should know.

These two answers are not different. Use the property of logarithms to rewrite $-\ln|\cos(x)|$ as $\ln|(\cos(x))^{-1}|$. This is equal to $\ln\left|\frac{1}{\cos(x)}\right|$, which in turn is equal to $\ln|\sec(x)|$.

BE AWARE

The expression $(\cos(x))^{-1}$ means the reciprocal of $\cos(x)$ while the expression $\cos^{-1}(x)$ is the inverse function for $\cos(x)$. They are not the same.

Example 12: Evaluate $\int \sec(x)\,dx$.

Solution: This one is downright tricky. The trick to this problem is to recognize that the derivatives of $\tan(x)$ and $\sec(x)$ both involve $\sec(x)$.

1. So the not so obvious process is to multiply $\sec(x)$ by the expression $\frac{\sec(x)+\tan(x)}{\sec(x)+\tan(x)}$, and the problem now becomes $\int \frac{\sec^2(x)+\sec(x)\tan(x)}{\sec(x)+\tan(x)}\,dx$.

2. Let $u = \sec(x) + \tan(x)$ and $du = (\sec(x)\tan(x) + \sec^2(x))\,dx$, and transform the problem to $\int \frac{1}{u}\,du = \ln|u| + C$.

3. Therefore, $\int \sec(x)\,dx = \ln|\sec(x) + \tan(x)| + C$.

CRITICAL POINT

Add $\int \sec(x)\,dx = \ln|\sec(x) + \tan(x)| + C$ to the list of important antiderivatives you should know.

Example 13: Evaluate $\int \tan^3(x)\,dx$.

Solution:

1. Rewrite $\tan^3(x)$ as $\tan(x)\tan^2(x)$ and use the trigonometric identity $\tan^2(x) = \sec^2(x) - 1$.

2. The integral is now $\int \tan(x)\big(\sec^2(x) - 1\big)dx =$
   $\int \tan(x)\sec^2(x) - \tan(x)\,dx = \frac{1}{2}\tan^2(x) - \ln|\sec(x)| + C$.

Fundamental Theorem of Calculus

The genius, which led from being able to find instantaneous rates of change to the area under a curve, is something that has rarely been matched in mathematics. How this was done is something we'll explore when we discuss numerical approximations for the integral later in this chapter.

There are actually two versions of the *Fundamental Theorem of Calculus.* The first is one that those who have studied calculus can easily remember.

> **DEFINITION**
>
> **Fundamental Theorem of Calculus** states if f(x) is a differentiable function on the interval (a,b), exists at $x = a$ and $x = b$, and has the property f(x) = F'(x),
> then $\int_a^b f\big(x\big)dx$ = F(b) − F(a).

For example, $\int_1^3 x^2\,dx = \frac{1}{3}x^3\Big|_1^3 = \frac{1}{3}(3)^3 - \frac{1}{3}(1)^3 = \frac{26}{3}$.

Example 14: Evaluate $\int_0^8 \sqrt{x+1}\,dx$.

Solution: The antiderivative for $\sqrt{x+1}$ is $\frac{2}{3}(x+1)^{3/2}$, so $\int_0^8 \sqrt{x+1}\,dx = \frac{2}{3}(x+1)^{3/2}\Big|_0^8 =$
$\frac{2}{3}(9)^{3/2} - \frac{2}{3}(1)^{3/2} = \frac{52}{3}$.

Let's look at a couple problems that appear to be more challenging. Remember, you are trying to match these problems against the list of basic integrals in the table at the beginning of this chapter.

Example 15: Determine the value of $\int_0^2 \frac{4x-3}{\sqrt{4x^2-6x+4}} dx$.

Solution: The integrand can be rewritten as $(4x - 3)(4x^2 - 6x + 4)^{-1/2}$. The "complicated" piece of the integrand is $(4x^2 - 6x + 4)^{-1/2}$.

1. Notice that the derivative of the base of this exponential statement is $8x - 6$ and that this is twice the value of the numerator.

2. Let $u = 4x^2 - 6x + 4$ so $du = (8x - 6)dx$ and $\frac{1}{2}du = (4x - 3)dx$.

3. Change the bounds of the integral—when $x = 0$, $u = 4(0)^2 - 6(0) + 4 = 4$, and when $x = 2$, $u = 4(2)^2 - 6(2) + 4 = 8$.

4. The problem $\int_0^2 \frac{4x-3}{\sqrt{4x^2-6x+4}} dx$ now becomes $\frac{1}{2}\int_4^8 \frac{1}{\sqrt{u}} du$.

5. The antiderivative of $\frac{1}{2}\left(\frac{1}{\sqrt{u}}\right)$ is $\sqrt{u}$.

6. Therefore, $\int_0^2 \frac{4x-3}{\sqrt{4x^2-6x+4}} dx = \frac{1}{2}\int_4^8 \frac{1}{\sqrt{u}} du = \sqrt{8} - \sqrt{4} = 2\sqrt{2} - 2$.

Example 16: Find the value of $\int_0^{\pi/6} \frac{1}{\sec(x)+\tan(x)} dx$.

Solution: This is just plain ugly! The numerator is not the derivative of the denominator. What to do? Some say "When in doubt, punt!" However, I say, when in doubt, going back to basics is a better approach.

1. Let's take the denominator and rewrite it in terms of sine and cosine, $\sec(x) + \tan(x) = \frac{1}{\cos(x)} + \frac{\sin(x)}{\cos(x)}$ and that equals $\frac{1+\sin(x)}{\cos(x)}$. That's better.

2. We can now rewrite the integrand $\frac{1}{\sec(x)+\tan(x)}$ as $\frac{\cos(x)}{1+\sin(x)}$. Now we have something with which we can work!

3. The numerator is the derivative of the denominator, so let $u = 1 + \sin(x)$ and $du = \cos(x)\,dx$.

4. The bounds of integration become 1 and 1.5 (when $x = 0$, $u = 1$ and when $x = \frac{\pi}{6}$, $u = 1.5$).

5. Evaluate the transformed integral:

$$\int_1^{1.5} \frac{1}{u} du = \ln|u|\Big|_1^{1.5} = \ln(1.5) - \ln(1) = \ln(1.5)$$

A graphing calculator can easily do these problems. For example, Example 15 becomes this:

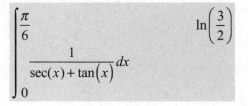

Figure 3.1
A screen shot from the TI-Nspire calculator evaluating the integral $\int_0^2 \frac{4x-3}{\sqrt{4x^2-6x+4}}dx$.

Example 16 becomes this:

$$\int_0^{\frac{\pi}{6}} \frac{1}{\sec(x)+\tan(x)}dx \qquad \ln\left(\frac{3}{2}\right)$$

Figure 3.2
A screen shot from the TI-Nspire calculator evaluating the integral $\int_0^{\pi/6} \frac{1}{\sec(x)+\tan(x)}dx$.

The question that always arises (and it's a legitimate question) is: "If the calculator can determine the answer, why do I need to learn this material?" The best answer I can give you is that there is a process to learning math. I cannot tell you every problem will lead to some deeper theory down the line, but the expectations at this point are that you will be able to recognize the pattern that the problem takes and can solve this problem. When I was in graduate school (oh, so long ago) taking a course in differential equations, we were allowed to bring a reference book to exams with us that contained about 1,000 integral patterns. You can be sure it was to our advantage to recognize the pattern of the problem we were solving so that we could find the correct integral.

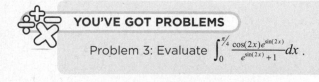

YOU'VE GOT PROBLEMS

Problem 3: Evaluate $\int_0^{\pi/4} \frac{\cos(2x)e^{\sin(2x)}}{e^{\sin(2x)}+1}dx$.

The *Second Fundamental Theorem of Calculus* states the nature of the inverse operations between the derivative and the integral, $\dfrac{d\left(\int_a^x f(t)\,dt\right)}{dx} = f(x)$. The derivative of the integral of a function is a function. Notice that the upper bound on the integral is the variable x, not a constant. The proof of this is fairly straightforward.

> **DEFINITION**
>
> The **Second Fundamental Theorem of Calculus** states that the derivative of an integral is the integrand.

The First Fundamental Theorem of Calculus tells us $\int_a^x f(t)\,dt = F(x) - F(a)$. The derivative of $F(x)$ is $F'(x)$ or $f(x)$ while the derivative of the constant $F(a)$ is 0.

Example 17: If $g(x) = \int_0^x \sin\left(\cos\left(t^3\right)\right)dt$, find g'(x).

Solution: According to the Second Fundamental Theorem of Calculus, the answer is $\sin(\cos(x^3))$.

Looks easy, doesn't it? What happens when things look too easy? Someone has to come along and make it harder. Let's look at Example 11 again but with one change made to the problem.

1. If $g(x) = \int_0^{\sqrt{x}} \sin\left(\cos\left(t^3\right)\right)dt$, find g'(x).

2. The upper bound is now a function rather than just the variable x.

3. If f(t) is $\sin(\cos(t^3))$, then $g(x) = \int_0^{\sqrt{x}} \sin\left(\cos\left(t^3\right)\right)dt = F\left(\sqrt{x}\right) - F(0)$.

4. Therefore, $g'(x) = F'\left(\sqrt{x}\right)\frac{1}{2\sqrt{x}} = \sin\left(\cos\left(x^{3/2}\right)\right)\frac{1}{2\sqrt{x}}$.

Example 18: Find f'(x) if $f(x) = \int_{\ln(x)}^9 \tan\left(e^t\right)dt$.

Solution: The Fundamental Theorem of Calculus says that $\int_{\ln(x)}^9 \tan\left(e^t\right)dt = G(9) - G(\ln(x))$ where $G(x)$ is the antiderivative of $\tan(e^t)$. Therefore, $f'(x) = G'(9) - G'(\ln(x))\frac{1}{x}$, so $f'(x) = \tan(e^{\ln(x)})\frac{1}{x} = \frac{\tan(x)}{x}$.

> **YOU'VE GOT PROBLEMS**
>
> Problem 4: Find f'(x) if $f(x) = \int_{\sin^{-1}(x)}^{e^{x^2}} \sin^3\left(t\right)dt$.

Numerical Approximation

The process of putting bounds on the integration and getting a specific value as the result is called a *definite integral*. The process of finding the antiderivative is called an *indefinite integral*. Before we get into using the definite integral, let's take a look at a few simple questions that one might see in Algebra I.

Example 19: In physics, work is defined as applying a force over a distance, and the work done is calculated by the product of the force applied and the distance involved. The accompanying diagram illustrates the amount of force applied and the distance involved. How much work is done?

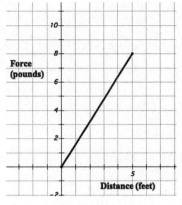

Figure 3.3

The diagram shows the amount of force applied over a given distance. At distance d = 0, there is no force being applied.

Solution: The amount of force increases at a steady rate. The amount of work done is represented by the triangle formed by this graph. Therefore, the amount of work done is $0.5 \times 8 \times 5 =$ 20 foot-pounds of work.

Example 20: Compute the amount of work done in the experiment, which has its results graphed in the following figure.

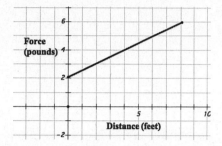

Figure 3.4
The diagram shows the amount of force applied over a given distance. At distance d = 0, there are 2 pounds of force being applied.

Solution: The region depicted under the line is a trapezoid. Therefore, the amount of work done is calculated as $0.5 \times 8 \times (2 + 6) = 32$ foot-pounds.

Example 21: The accompanying graph represents the velocity of a car for a portion of a trip. How far did the car travel during this time?

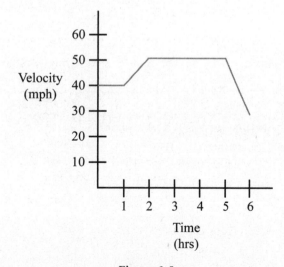

Figure 3.5
The diagram shows the speed a car is traveling during the time interval 0 to 6 hours.

Solution:

1. We'll start this problem by drawing some lines from the corners of the graph down to the horizontal axis.

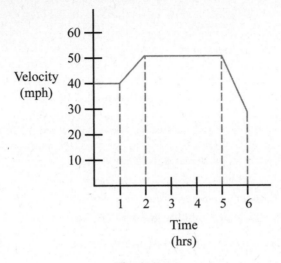

Figure 3.6

Partitioning the diagram at convenient intervals allows us to calculate the distance traveled during this 6-hour period.

2. The car travelled at 40 mph for the first hour for a total of 40 miles.

3. For the next hour, the driver gradually increased speed from 40 to 50 mph.

4. Compute the area of the trapezoid to determine the distance travelled during this time frame: $\frac{1}{2}(1)(40 + 50) = 45$ miles.

5. The car travels for 3 hours at 50 mph for a total of 150 miles.

6. We use the trapezoid to find the distance travelled during the last hour: $\frac{1}{2}(1)(30 + 50) = 40$ miles.

7. The car traveled a total of 275 miles during this time.

CRITICAL POINT

The area for each region is the product of miles per hour and hours giving a result of miles. This is a big piece of how the definite integral is applied.

Example 22: The graph of $y = f(x)$ is drawn here on the interval $[-10,10]$ and $G(x) = \int_{-4}^{x} f(x)\,dx$.

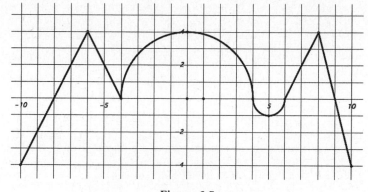

Figure 3.7
The graph consists of line segments and semicircles.

The graph of f(x) consists of line segments and semicircles.

1. Evaluate: (a) G(0) (b) G(4) (c) G(6) (d) G(10) (e) G(–10)

2. Determine the intervals when G(x) is increasing/decreasing.

3. Determine the relative maxima and relative minima of G(x).

4. Determine the concavity and the points of inflection for G(x).

Solution:

1. (a) $G(0) = \int_{-4}^{0} f(x)\,dx$. This is a quarter of a circle with a radius of 4, so $G(0) = 4\pi$.

   (b) $G(4) = \int_{-4}^{4} f(x)\,dx$. This is the area of the large semicircle with a radius of 4, so g(4) = 8π.

   (c) $G(6) = \int_{-4}^{6} f(x)\,dx = \int_{-4}^{4} f(x)\,dx + \int_{4}^{6} f(x)\,dx$. The area denoted by $\int_{4}^{6} f(x)\,dx$ is a semicircle with a radius of 1, which lies below the x-axis and has negative area. Therefore, $G(6) = \frac{15\pi}{2}$.

(d) $G(10) = \int_{-4}^{10} f(x)\,dx = \int_{-4}^{6} f(x)\,dx + \int_{6}^{9} f(x)\,dx + \int_{9}^{10} f(x)\,dx \cdot \int_{6}^{9} f(x)\,dx$

represents the area of the triangle that lies above the x-axis. The area of this triangle is 6.

$\int_{9}^{10} f(x)\,dx$ represents the area of the right triangle that would be formed by dropping a perpendicular line from $(10, -4)$ to $(10, 0)$. Because this triangle lies below the x-axis, the area is considered negative and $\int_{9}^{10} f(x)\,dx = -2$. Therefore, $G(10) = \frac{15\pi}{2} + 4$.

(e) $G(-10) = \int_{-4}^{-10} f(x)\,dx = -\int_{-10}^{-4} f(x)\,dx$. There are two triangles that make up this interval. The triangle above the x-axis has an area of 8, and the area of the triangle below the x-axis is -4. Therefore, $\int_{-4}^{-10} f(x)\,dx = -(-4 + 8) = -4$.

2. To determine when $G(x)$ is increasing or decreasing, we need to examine the signs of $G'(x)$. Because $G(x) = \int_{-4}^{x} f(x)\,dx$, then by the Second Fundamental Theorem of Calculus $G'(x) = f(x)$. $G'(x) = 0$ when $f(x) = 0$ and that occurs when $x = -8, -4, 4, 6,$ and 9. $G'(x) < 0$ on the intervals $[-10, -8)$, $(4,6)$, and $(9, 10]$ because $f(x) < 0$ on those intervals. These are the intervals when $g(x)$ is decreasing. $G'(x) > 0$ on $(-8,-4)$, $(-4,4)$, and $(6,9)$. These are the intervals when $G(x)$ is increasing.

3. $G(x)$ has relative minima whenever G goes from a decreasing function to an increasing function. This happens when $x = -8$ and 6. $G(-8) = -8$ (see the argument from (1e) earlier) and $G(6) = \frac{15\pi}{2}$. $G(x)$ has relative maxima when G goes from an increasing function to a decreasing function. This happens when $x = 4$ and $x = 9$. So $G(4) = 8\pi$ and $G(9) = \frac{15\pi}{2} + 6$.

4. To determine the points of inflection for $G(x)$ we need to examine $G''(x) = f'(x)$. $G''(x) = 0$ at $x = 0$ and $x = 5$. $G''(x)$ fails to exist at $x = -6$, $x = -4, 4, 6,$ and 8. $G''(x) > 0$ on $[-10, -6)$, $(-4, 0)$, $(5,6)$, and $(6,8)$. Therefore, $G(x)$ is concave up for these intervals. $G''(x) < 0$ on $(-6, -4)$, $(0, 4)$, $(4,5)$, and $(8, 10]$. $G(x)$ is concave down on these intervals. Therefore, the points of inflection occur when $x = -6$, $x = -4$, $x = 0$, $x = 5$, and $x = 8$. The points of inflection occur at $G(-6) = -4$, $G(-4) = 0$, $G(0) = 4\pi$, $G(5) = \frac{15\pi}{2}$, and $G(8) = \frac{15\pi}{2} + 4$.

The Least You Need to Know

- Basic integration formulas let you determine the function whose derivative is given to you in the integrand.

- The First Fundamental Theorem of Calculus enables you to compute the area under a curve. Very often this value has an application beyond just finding the area.

- The Second Fundamental Theorem of Calculus reinforces the notion that derivatives and integrals are inverse operations and also allows for the analysis of functions that are defined by an integral.

- Use u-substitution when dealing with complicated integrands that are the form of the basic integration formulas.

- The area under a graph can represent physical properties.

Length, Area, and Volumes

Calculus is really the study of two items—measuring how fast something is changing and accumulating material a little bit at a time. You've had a fair amount of experience with the measuring rates of change. In Part 2, we take some time to work on the accumulation. Keep in mind, we are still looking at mathematics through the microscope. That means what might not look like much of a region to the naked eye can be rather formidable when we zoom in on it.

You're asked to use your imagination when we look at volumes. First, we begin working in a two-dimensional plane but then we rotate figures around vertical and horizontal lines. You then are asked to stand "at the end of the line" and look at a cross section. (Think of cutting a piece of fruit with a sharp knife and turning what was an interior part of the fruit toward you.) Getting the vision of the cross section in your mind goes a long way to helping you work with volumes.

The Pythagorean Theorem also comes into play again as you use it in very small increments to measure the length along a curve. You also learn about the area formulas for rectangles and trapezoids used to help find areas of regions with curved boundaries.

Areas and Approximations

As you read in Chapter 3, the area under a curve can have meaningful applications. While the problems in Chapter 3 consisted of functions that were linear in nature—or composed of line segments—it is also important to be able to compute the area under curved boundaries.

In This Chapter

- Using plane geometry to approximate the area under a curve

- Computing the true area under a curve

- Average value of a function

- Computing the true area between curves

- Approximate the area under a curve with parabolic arcs

Riemann Sums

We know that the area of a rectangle is equal to the product of length and width, and in this context, the length would be a functional value. That means we can estimate the area of a region by finding a reasonable point to measure the height and use the width of the interval as the width of the rectangle. Furthermore, if we remember that one of the ideas of calculus is that we are putting algebra under a microscope, we can subdivide the interval into as many sections as we want. We can also make those sections as small as we like. The idea is that the total area under the curve is the sum of the areas of all these subsections, regardless of how many sections or the size of those sections. The key is to find a "reasonable" point to measure height.

> **CRITICAL POINT**
>
> Bernhard Riemann, a nineteenth-century German mathematician, is credited with examining this process. Consequently, the estimation of the bounded integral is called a Riemann Sum.

Let's start off with the basics. We'll take the function $y = f(x)$ on the interval $[a,b]$ and require that $f(x)$ be continuous and that $f(x) > 0$ throughout this interval. We'll call the region bounded by $y = f(x)$, $x = a$, $x = b$, and the x-axis R. We could call it anything we want but R is the first letter in region so that seems simple enough.

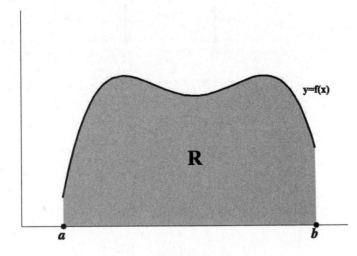

Figure 4.1
R is the region bounded by the graph of $y = f(x)$, the x-axis, $x = a$, and $x = b$.

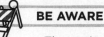

BE AWARE

The notion that the function be non-negative allows for an easy generalization of the theory. If there is a section or two that falls below the x-axis, they could be handled by choosing appropriate partition points where the graph intersects the x-axis. Then we would subtract the area for these regions from the sum of the regions that lie above the axis.

We'll divide the region into n partitions at the points $x_1, x_2, x_3, \ldots, x_{n-1}$, and x_n. The width of the partitions does not need to be the same size. Each width will be designated as Δx_i. The width $\Delta x_1 = x_1 - a$, $\Delta x_2 = x_2 - x_1$, $\Delta x_3 = x_3 - x_2$, and $\Delta x_n = b - x_{n-1}$.

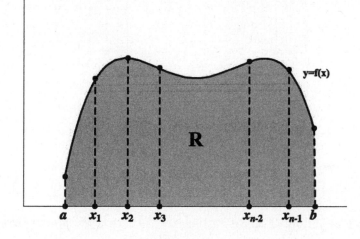

Figure 4.2
R is partitioned at the points $x_0 = a, x_1, x_2, x_3, \ldots x_{n-2}, x_{n-1}, x_{n=b}$.

For each partition, choose a point c_i and draw a segment from $(c_i, 0)$ to $(c_i, f(c_i))$. For each partition, construct a rectangle with width Δx_i and height $f(c_i)$.

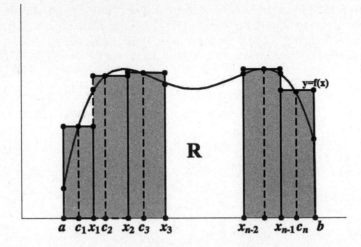

Figure 4.3

Rectangles are drawn using the midpoint of each partition to determine the height of the rectangle.

The estimate for the area under the curve is found by finding the sum of the areas in all the rectangles. Clearly, the more rectangles formed, the better the estimate for the true area becomes. The true area of the region under the curve can be defined as $A = \lim\limits_{\max \Delta x_i \to 0} \sum\limits_{i=1}^{n} f(c_i)\Delta x_i$.

The points c_i can be anywhere within the interval $[x_{i-1}, x_i]$. Three "easy" points come immediately to mind:

- The left endpoint

- The right endpoint

- The midpoint of each interval

A fourth method for estimating the area under the curve is to use trapezoids rather than rectangles.

Example 1: Consider the function $f(x) = x^3 - 3x^2 + 6$. We'll look at the interval $[0, 2]$ and subdivide into 10 intervals and use each of these methods to estimate the area under the curve in this interval.

Solution: The width of each interval is $\frac{2-0}{10} = 0.2$.

Left endpoint: From each of the values 0, 0.2, 0.4, 0.6, …, 1.8, draw perpendicular segments to the graph of $y = f(x)$. Using the left endpoint as the height of a rectangle, draw the rectangle with width 0.2 units.

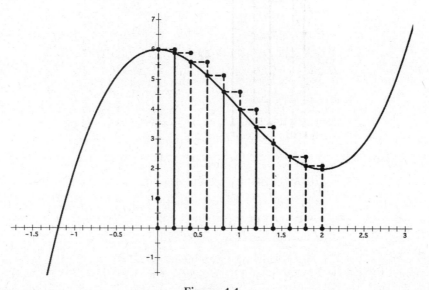

Figure 4.4
Rectangles are drawn using the left endpoint of each partition to determine the height of the rectangle.

The approximate area under the curve is the sum of these rectangles. You can use the List feature on your calculator to compute this sum. In one list, enter each of the values of x. They can be entered manually or you can use the SEQ command on your calculator to generate them. In this case, the command would be seq(x,x,0,1.8,0.2). Compute the corresponding functional values in the second list. Finally, find the sum of the second list and multiply this sum by the width of the interval to get the result.

CRITICAL POINT

The SEQ command uses the inputs seq(expression, independent variable, starting value, ending value, increment). In this example, the expression is simply $y = x$ and the increment is 0.2. This is true because the width of the interval divided by the 10 intervals is 0.2.

=	=seq(x,x,0,1.8,0.2)	=f(in)
1	0	6
2	0.2	5.888
3	0.4	5.584
4	0.6	5.136
5	0.8	4.592
6	1.	4.
7	1.2	3.408
8	1.4	2.864
9	1.6	2.416
10	1.8	2.112

Figure 4.5
The spreadsheet display from the TI-Nspire shows the coordinate of the left endpoint of the partition and the functional value of that point (which is the height of the rectangle for the partition).

I try not to get too caught up with labeling the columns on the TI-Nspire when I am using the spreadsheet feature. I tend to call my input values *in* and my output values *out*. They are not very original but they also do not require a lot of typing. And to be very creative, if I need to use another set of input and output values in the same spreadsheet, I tend to call them *in1* and *out1*.

The result is 0.2 times the sum of the column *Out*, 8.4. Clearly, each rectangle in this example has more area within it than does the corresponding interval bounded on top by the graph.

Right endpoint: The process is essentially the same except the rectangles are drawn from the right endpoint of each interval.

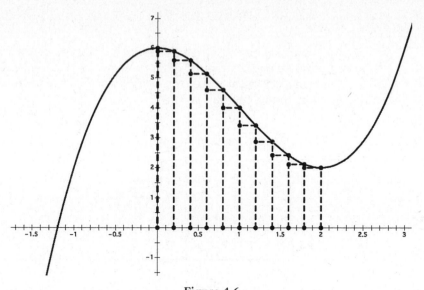

Figure 4.6
Rectangles are drawn using the right endpoint of each partition to determine the height of the rectangle.

The table of *x* values will run from 0.2 to 2.0.

=	=seq(x,x,0.2,2,0.2	=f(in)
1	0.2	5.888
2	0.4	5.584
3	0.6	5.136
4	0.8	4.592
5	1.	4.
6	1.2	3.408
7	1.4	2.864
8	1.6	2.416
9	1.8	2.112
10	2.	2.

Figure 4.7
The spreadsheet display from the TI-Nspire shows the coordinate of the right endpoint of the partition and the functional value of that point (which is the height of the rectangle for the partition).

The sum of the areas for this region is 7.6. Clearly, this estimate is less than the true area because there is a gap between the rectangle and the graph above it.

Midpoint: Use the midpoint of each interval to determine the height of each rectangle.

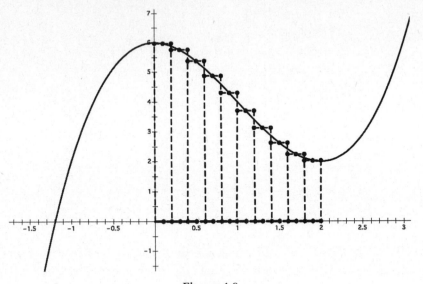

Figure 4.8

Rectangles are drawn using the midpoint of each partition to determine the height of the rectangle.

The values of x now run 0.1, 0.3, 0.5, …, 1.9.

A in	B out	C	D	E	F
=seq(x,x,0.1,1.9,0..	=f(in)				
0.1	5.971				
0.3	5.757				
0.5	5.375				
0.7	4.873				
0.9	4.299				
1.1	3.701				
1.3	3.127				
1.5	2.625				
1.7	2.243				
1.9	2.029				

Figure 4.9

The spreadsheet display from the TI-Nspire shows the coordinate of the midpoint of the partition, and the functional value of that point (which is the rectangle's height for the partition).

The sum of the areas of these rectangles is 8. Each rectangle appears to have some area above the graph. Also missing is some area under the graph of $y = f(x)$.

Trapezoid: The height of each trapezoid is the width of the interval, while the bases of the trapezoid are the functional values at each endpoint.

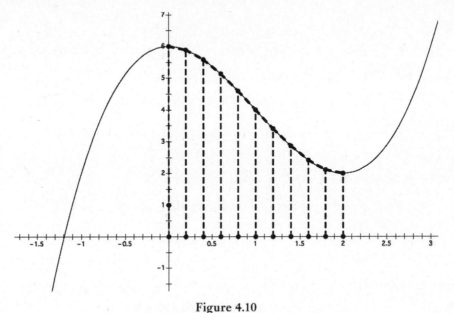

Figure 4.10

Trapezoids are drawn using the endpoints of the partitions. Notice that with the exception of x = a and x = b, all partition points are used twice.

The values of x now run from 0 to 2 in increments of 0.2 and include both endpoints.

=	=seq(x,x,0,2,0.2)	=f(in)
1	0	6
2	0.2	5.888
3	0.4	5.584
4	0.6	5.136
5	0.8	4.592
6	1.	4.
7	1.2	3.408
8	1.4	2.864
9	1.6	2.416
10	1.8	2.112
11	2	2

Figure 4.11

The spreadsheet display from the TI-Nspire shows the coordinate of each partition point and the functional value of that point. (These values are the lengths of the bases for the partition.)

Recall that the formula for the area of a trapezoid is $\frac{1}{2}h(b_1 + b_2)$. The trapezoidal estimate is

$\frac{1}{2}(0.2)(6+5.888) + \frac{1}{2}(0.2)(5.888+5.584) + \frac{1}{2}(0.2)(5.884+5.136) + ... + \frac{1}{2}(0.2)(2.416+2.112) + \frac{1}{2}(0.2)(2.112+2)$

$= 8$.

Notice how this equation becomes $\frac{1}{2}(0.2)(f(0) + 2f(0.2) + 2f(0.4) + ... + 2f(1.8) + f(2))$.

We'll find the true area under this curve in a few moments.

An estimate found using the trapezoidal method will always be the average of the left endpoint and right endpoint methods.

YOU'VE GOT PROBLEMS

Problem 1: Use the midpoint method with 20 subdivisions to estimate $\int_1^5 x^4 - 5x^2 + 9\,dx$.

Example 2: Approximate the area under the curve $g(x) = x^2\sqrt{25 - x^2}$ on the interval [1, 4] with 20 partitions using each of the four methods discussed.

Solution: The width of each interval is $\frac{4-1}{20} = 0.15$.

A look at the graph shows that it is continuous and non-negative on the interval [1, 4].

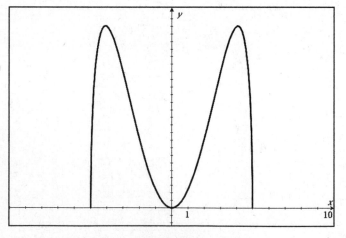

Figure 4.12
The graph of $g(x) = x^2\sqrt{25 - x^2}$ *on the interval [–5, 5].*

Left endpoints: The left endpoints for the partitions will be 1, 1.15, 1.30, 1.45, ..., 3.70, 3.85 and can be generated with the Sequence command seq($x, x,$ 1, 3.85, 0.15). The heights for the rectangles will be calculated at these points.

◆	A in	B out
=	=seq(x,x,1,3.85,0.15)	=g(in)
1	1	2*√(6)
2	1.15	6.43522
3	1.3	8.15939
4	1.45	10.0607
5	1.6	12.1269
6	1.75	14.344
7	1.9	16.696
8	2.05	19.1652
9	2.2	21.7315
10	2.35	24.3726

Figure 4.13
A partial spreadsheet display from the TI-Nspire shows the coordinate of the left endpoint of the partition and the functional value of that point.

The estimate for the area of the specified region is 0.15 times the sum of these values. Use this command on a calculator: The 0.15*sum(out) = 78.0529.

Right endpoints: The heights for the rectangles will be calculated when x = 1.15, 1.30, 1.45, ..., 3.85, and 4.

◆	A in	B out
=	=seq(x,x,1.15,4,0.1	=g(in)
1	1.15	6.43522
2	1.3	8.15939
3	1.45	10.0607
4	1.6	12.1269
5	1.75	14.344
6	1.9	16.696
7	2.05	19.1652
8	2.2	21.7315
9	2.35	24.3726

Figure 4.14
A partial spreadsheet display from the TI-Nspire shows the coordinate of the partition's right endpoint and the functional value of that point.

The estimate for the area using right endpoint partitions is 85.518.

Trapezoid: The trapezoidal approximation is the average of the left and right endpoint methods. This means the approximation is $\frac{78.0529 + 85.518}{2} = 81.2855$.

Midpoints: The heights of the rectangles will be calculated when $x = 1.075, 1.225, 1.375, \ldots$, and 3.925.

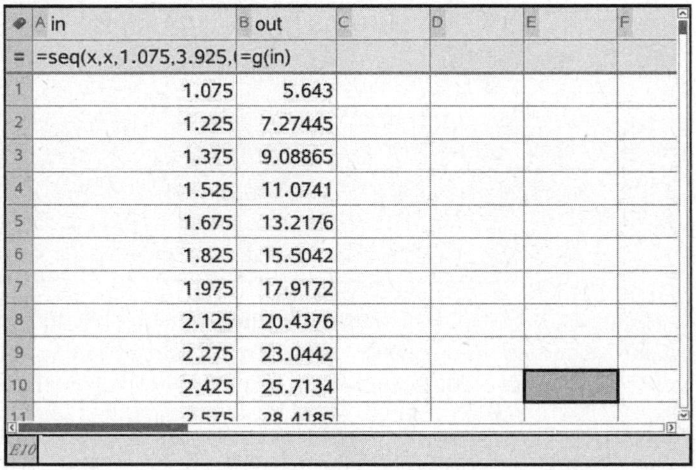

◆	A in	B out	C	D	E	F
=	=seq(x,x,1.075,3.925,(	=g(in)				
1	1.075	5.643				
2	1.225	7.27445				
3	1.375	9.08865				
4	1.525	11.0741				
5	1.675	13.2176				
6	1.825	15.5042				
7	1.975	17.9172				
8	2.125	20.4376				
9	2.275	23.0442				
10	2.425	25.7134				
11	2.575	28.4185				

E10

Figure 4.15

A partial spreadsheet display from the TI-Nspire shows the coordinate of the partition's midpoint and the functional value of that point.

The estimate for the area using midpoint approximations is 81.3049.

Example 3: A plane flies in a straight line with positive velocity $v(t)$ kilometers per minute with v being a differentiable function of t. The following table shows the velocity of the plane during the interval [0, 60] in 5-minute increments.

t	$v(t)$
0	7.9
5	8.1
10	8.4
15	8.8
20	8.5
25	8.3
30	8.7
35	9.0
40	8.8
45	8.6
50	8.4
55	8.2
60	8.1

Use the trapezoidal method with 6 intervals to compute the Riemann Sum approximation for $\int_0^{60} v(t)\,dt$. Then explain the meaning of this result in terms of the plane's flight.

Solution: The width of each interval is 10 minutes so the Riemann Sum approximation for $\int_0^{60} v(t)\,dt$ is $\frac{1}{2}(10)\big(v(0)+2v(10)+2v(20)+2v(30)+2v(40)+2v(50)+v(60)\big) =$

$\frac{1}{2}(10)\big(7.9+2(8.4+8.5+8.7+8.8+8.4)+8.1\big) = 508$ kilometers traveled in 60 minutes.

From Numerical Approximations to the True Area

The Mean Value Theorem states that under the conditions that f(x) is differentiable on the interval (a,b) and continuous on [a,b], there will be a value of c in the interval so that

$f'(c) = \frac{f(b)-f(a)}{b-a}$. If we multiply both sides of the equation by the denominator, then f'(c) (b – a) = f(b) – f(a). Let's change the notation from f'(c) and f(c) to represent the derivative and antiderivative to f(c) and F'(c). The equation from the Mean Value Theorem is now f(c)(b – a) = F(b) – F(a). This is where the true genius of calculus is found. If we look at the approximations of the area using the various values for c_i, the area for each partition is:

$f(c_1)*\Delta x_1 = F(x_1) - F(a)$

$f(c_2)*\Delta x_2 = F(x_2) - F(x_1)$

$f(c_3)*\Delta x_3 = F(x_3) - F(x_2)$

$f(c_4)*\Delta x_4 = F(x_4) - F(x_4)$

And so on until:

$f(c_n)*\Delta x_n = F(b) - F(x_{n-1})$

The area is the sum of the $f(c_i)*\Delta x_i$, which in turn equals:

$F(x_1) - F(a) + F(x_2) - F(x_1) + F(x_3) - F(x_2) + \dots + F(b) - F(x_{n-1})$

This collapses to F(b) – F(a) = $\int_a^b f(x)\,dx$.

How cool is that!

The true area under the curve $y = x^3 - 3x^2 + 6$ on the interval [0, 2] is $\int_0^2 x^3 - 3x^2 + 6\,dx$ = 8.

(We were fortunate to find the exact area with 10 subdivisions. That will rarely happen. But the estimate that you will get should be pretty close to 10 subdivisions.)

At this point in our study of integrals, we do not yet know the antiderivative for $g(x) = x^2\sqrt{25 - x^2}$, so we will have to be satisfied with our approximations for the time being— unless you want to let the calculator do the work for you. We all know that you do, so here it is:

$$\int_1^4 g(x)\,dx \qquad \frac{625 \cdot \cos^{-1}\left(\frac{3}{5}\right)}{8} - \frac{625 \cdot \sin^{-1}\left(\frac{1}{5}\right)}{8} + \frac{23 \cdot \sqrt{6}}{4} + \frac{21}{2}$$

Figure 4.16
A screen shot from the TI-Nspire CAS calculator shows the exact value of the area of the region bounded by $y = g(x)$ and the x-axis from $x = 1$ to $x = 4$.

I'll bet you weren't expecting that answer! This is just a preview for how interesting determining integrals will get. Numerically, this is equal to 81.2984. The trapezoid and midpoint approximations were pretty close.

Example 4: Find the area under the first arch of the sine curve in the first quadrant.

Solution: The graph of the function $y = \sin(x)$ crosses the x-axis at $x = 0$ and $x = \pi$. The area under the curve is found using the definite integral $\int_0^\pi \sin(x)\,dx = -\cos(x)\Big|_0^\pi = -\cos(\pi) - -\cos(0) = -(-1) + 1 = 2$.

Unlike all the work you did in geometry, you now have to deal with the fact that area is a vector rather than a scalar. That is, area can be negative.

Example 5: Evaluate $\int_\pi^{2\pi} \sin(x)\,dx$.

Solution: $\int_\pi^{2\pi} \sin(x)\,dx = -\cos(x)\Big|_\pi^{2\pi} = -\cos(2\pi) - -\cos(\pi) = -(1) - 1 = -2$.

Aside from the fact that this is the way the numbers work, why would we want to assign a direction to area? Rather than think of the problem in the sterile format of evaluating the integral, put the problem in context. For example, suppose the integrand is the velocity of an object and the goal is to find the displacement (not distance) of the object. In Example 4, the object will have moved 2 units in the positive direction (however positive is defined) while the object will have moved 2 units in the reverse direction putting it back in its original position. How far did the object travel? 4 units. What is its displacement? 0 units.

Example 6: Find the total displacement and total distance traveled by an object that moves along a horizontal line with velocity $v(t) = \sin\left(\frac{\pi t}{3}\right) + \cos\left(\frac{\pi t}{6}\right)$ for $0 \leq t \leq 9$.

Solution: The displacement is found using the definite integral $\int_0^9 \sin\left(\frac{\pi t}{3}\right) + \cos\left(\frac{\pi t}{6}\right) dt =$

$\frac{-3}{\pi}\cos\left(\frac{\pi t}{3}\right) + \frac{6}{\pi}\sin\left(\frac{\pi t}{6}\right)\Big|_0^9 = \left(\frac{-3}{\pi}\cos\left(3\pi\right) + \frac{6}{\pi}\sin\left(\frac{3\pi}{2}\right)\right) - \left(\frac{-3}{\pi}\cos\left(0\right) + \frac{6}{\pi}\sin\left(0\right)\right) =$

$\left(\frac{-3}{\pi}(-1) + \frac{6}{\pi}(-1)\right) - \left(\frac{-3}{\pi} + 0\right) = 0.$

We need to take a look at the velocity function to see if it ever reverses itself. This will cause the distance traveled to be different from the displacement.

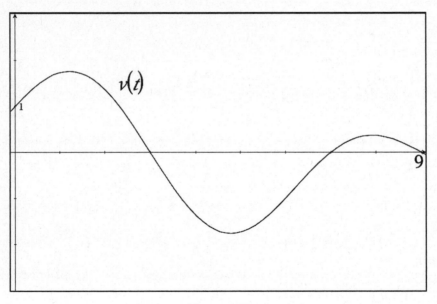

Figure 4.17
The graph of the velocity function $v(t) = \sin\left(\frac{\pi t}{3}\right) + \cos\left(\frac{\pi t}{6}\right)$ for $0 \leq t \leq 9$.

1. Let the calculator determine the points where the graph crosses the *t*-axis (also, where the object changes directions).

2. Use the zero function on the calculator to bind the first point where the graph crosses the *x*-axis ((3,0)) and then do the same for the second point of intersection ((7,0)).

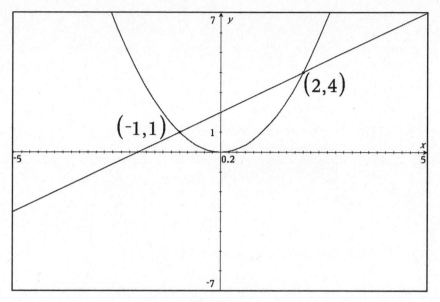

Figure 4.18
The velocity of the particle is 0 when t = 3, 7, and 9.

3. To find the total distance traveled, we need to calculate the definite integral from $t = 0$ to $t = 3$ and from $t = 7$ to $t = 9$, add these together, calculate the definite from $t = 3$ to $t = 7$, and subtract this value from the sum.

4. Let's be practical and let the calculator do the work for us. The total distance traveled is $\int_0^3 v(t)\,dt + \int_7^9 v(t)\,dt - \int_3^7 v(t)\,dt = \frac{27}{\pi}$.

YOU'VE GOT PROBLEMS

Problem 2: A particle moves along a straight line with velocity $v(t) = 4\cos\left(\frac{\pi t}{4}\right) + 2\sin\left(\frac{\pi t}{3}\right)$ during the interval [0,12]. Find the total displacement and total distance traveled by the object.

Average Value of a Function

You saw in Example 4 that the area under the first arch of the sine function is 2. What is the *average value* of the sine function on the interval [0, π]? Unlike days of old when one found an average by adding a set of numbers and then dividing by the number of values in the data set, we need to work with a continuous set of data. Fortunately, this turns out to be a fairly simple problem.

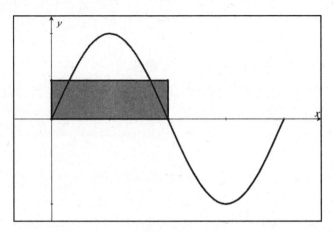

Figure 4.19

The graph of y = sin (x) on [0, 2π] and a rectangle illustrating the area under the sine curve from 0 to π.

The rectangle is drawn on the same interval as is the arch of the sine wave. The task is to now find the height of the rectangle that will give the same area as the area under the arch. We know the area and we know the width of the rectangle so all we need to do is divide these values.

> **DEFINITION**
>
> The **average value** of a continuous function f(x) on the interval [a, b] is
>
> $\frac{1}{b-a}\int_a^b f(x)\,dx$.

The average value of the sine function on [0, π] is $\frac{1}{\pi-0}\int_0^{\pi} \sin(x)\,dx = \frac{2}{\pi}$.

Example 7: Find the average value of the function $f(x) = \sqrt{x}$ on the interval [1, 9].

Solution: The average value is $\frac{1}{9-1}\int_1^9 \sqrt{x}\,dx = \frac{13}{6}$.

Example 8: When studying the depth of a bay due to the changes in tide, the model $d(t) =$ $10 + 3\cos\left(\frac{t}{12}\right) + 2\sin\left(\frac{t}{12}\right)$ is used, where d is measured in feet and t in hours. Determine the average value of $d(t)$ over the interval $[0, 12]$.

Solution: The average depth of the water is $\frac{1}{12}\int_0^{12} 10 + 3\cos\left(\frac{t}{12}\right) + 2\sin\left(\frac{t}{12}\right)dt =$ $\left(\frac{1}{12}\right)\left(10t + 36\cos\left(\frac{t}{12}\right) + 24\sin\left(\left(\frac{t}{12}\right)\right)\right)\Big|_0^{12}$ $\left(\frac{1}{12}\right)\left(10t + 36\sin\left(\frac{t}{12}\right) + 24\cos\left(\left(\frac{t}{12}\right)\right)\right)\Big|_0^{12} = 13.44$ feet.

Area Between Two Curves

The definite integral can also be used to find the area between two curves. As you did in geometry, you take the area of the larger figure and subtract the area of the smaller figure. In the application of the definite integral, you need to determine which graph is above the other in the interval involved.

Example 9: Find the area of the region bounded by the line $y = x + 2$ and the parabola $y = x^2$.

Solution: Use your graphing utility to look at a picture of the graphs.

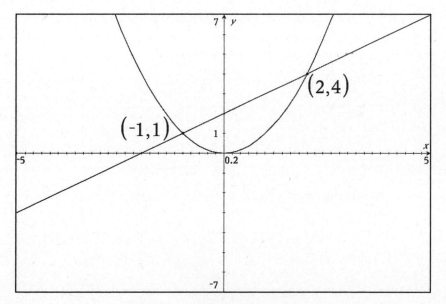

Figure 4.20
The graphs of $y = x + 2$ and $y = x^2$ intersect at $(-1, 1)$, and $(2, 4)$.

The graphs intersect at $x = -1$ and $x = 2$. During this interval, the line forms the upper boundary for the region so the area is found using the integral $\int_{-1}^{2}(x+2)-\left(x^2\right)dx = \frac{1}{2}x^2 + 2x - \frac{1}{3}x^3\Big|_{-1}^{2} = $

$\left(\frac{1}{2}(2)^2 + 2(2) - \frac{1}{3}(2)^3\right) - \left(\frac{1}{2}(-1)^2 + 2(-1) - \frac{1}{3}(-1)^3\right) = \left(2 + 4 - \frac{8}{3}\right) - \left(\frac{1}{2} - 2 - \frac{1}{3}\right) = \frac{9}{2}$.

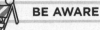

BE AWARE

When computing the area between two curves, the difference in the functions is always the function with greater values on the interval minus the function with lesser values. If reversed, the area will be a negative value.

Example 10: Find the total area of the regions bounded by the graphs of $y = x^2$ and $y = 2^x$.

Solution: This is an interesting problem for a few reasons.

First, you have a parabola and an exponential intersecting and there is no clean algebraic method for determining the points of intersection. *Ah yes*, you say to yourself, *go get the graphing calculator.* When you stop to think about it, there are two obvious points of intersection, (2,4) and (4, 16). This is usually where most people stop thinking about the points of intersection and set up an integral to get the answer.

But recall that as x goes to negative infinity, the parabola grows to infinity while the exponential goes to zero. There must be a point somewhere to the left of $x = 0$ where the graphs intersect. This is where the calculator becomes very important.

(The other piece of this is that people often forget to think about the viewing window when graphing functions. Although they see the intersection to the left of zero, they ignore what happens beyond $y = 10$ when the graphs intersect a third time. The moral of this story is that you should not rely on the calculator to pick the appropriate viewing window for you. Instead, you must consider the behavior of the function. In this case, the exponential function grows faster than the parabola so the graphs must cross a third time.)

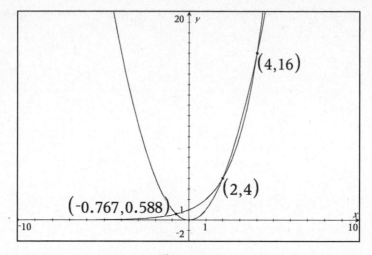

Figure 4.21
The graphs of $y = x^2$ and $y = 2^x$ intersect at three points.

1. Use the intersect feature from the calculator to determine the points of intersection, particularly the leftmost.

2. You can have the calculator store this in a variable so that you do not need to type all the decimal values when you are integrating. In this example, I stored the smallest value of x in the variable **a**.

3. The exponential function forms the upper boundary for the first region while the parabola does so for the second region.

4. Therefore, the total area bounded by these two functions will be the sum of two definite integrals.

$$\text{Area} = \int_a^2 2^x - x^2 \, dx + \int_2^4 x^2 - 2^x \, dx = 3.46025$$

YOU'VE GOT PROBLEMS

Problem 3: Find the total area of the regions bounded by $f(x) = x^3 - 3x^2 - 5x$ and $g(x) = 2x - 10$.

Example 11: Find the area of the region bounded in the first quadrant by the graphs of $y = \cos(x)$, $y = \sin(x)$, and $x = 0$.

Solution: The graph of this region is:

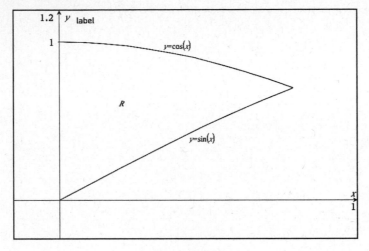

Figure 4.22

The region R is bounded in the first quadrant by the y-axis and the graphs of $y = \cos(x)$ and $y = \sin(x)$.

The two graphs intersect when $x = \frac{\pi}{4}$. The area between the curves is found with the integral

$$\int_0^{\pi/4} \cos(x) - \sin(x)\,dx = \sqrt{2} - 1.$$

Example 12: Find the area of the region bounded by the graphs of $y = \cos(2x)$ and $y = 0.5x$.

Use your graphing calculator to determine where these functions intersect.

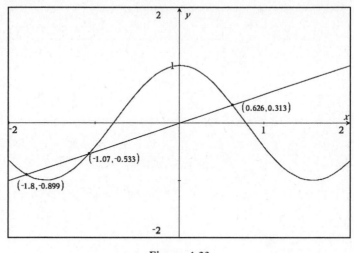

Figure 4.23
The graphs of $y = \cos(2x)$ and $y = 0.5x$ intersect at three points.

We'll need to use the technology to find the points of intersection because we have no way to do so algebraically. I stored the x-coordinate for the first point on intersection in variable a, the second in b, and the third in c.

Between a and b, the line forms the upper boundary while the graph of $y = \cos(2x)$ forms the upper boundary from b to c. The total area of these two regions is

$$\int_a^b 0.5x - \cos(2x)\,dx + \int_b^c \cos(2x) - 0.5x\,dx = 1.2029.$$

YOU'VE GOT PROBLEMS

Problem 4: Find the area bounded by the graphs of $y = \sec(x)$ and $y = 4 - x^2$, as shown in the accompanying diagram.

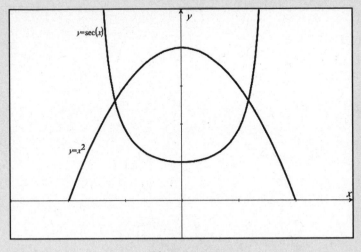

Figure 4.24
The graph shows the region formed by the intersection of $y = x^2$ and
$y = \sec(x)$.

Simpson's Rule

The techniques we studied earlier in this chapter for approximating the area under a curve all relied on plane geometry—that is, on boundaries that involved line segments. Simpson's Rule uses the Fundamental Theorem of Calculus to form an estimate for the area. "What's that?" you say. "You're going to use the Fundamental Theorem of Calculus to approximate the area under a curve when the Fundamental Theorem of Calculus can find the area under a curve. Explain that to me." Thank you, I will.

As we saw earlier, there are some functions that we do not yet know how to integrate, such as $g(x) = x^2\sqrt{25 - x^2}$. There are some functions, such as $y = e^{-x^2}$ that we will never be able to integrate and yet the values generated are important. *Simpson's Rule* divides the interval of integration into an *even* number of partitions. Three successive points generated by the endpoints over two partitions are used to generate a parabolic arc (if you are unfamiliar with this at this time, it is a technique called quadratic regression that can be used to generate a quadratic function that will pass through these three points and can be done with the regression feature on your graphing calculator). The Fundamental Theorem of Calculus is used to compute the area under this arc. The process is repeated until the entire interval is used.

📖 **DEFINITION**

> **Simpson's Rule** states you use the area under a parabolic arc determined by three data points to approximate the area under a curve.

The consequence of finding the quadratic equation and the Fundamental Theorem of Calculus is that the area under a particular arc from x_i to x_{i+2} is equal to $\frac{\Delta x}{3}\left(f\left(x_i\right) + 4f\left(x_{i+1}\right) + f\left(x_{i+2}\right)\right)$.

The results of this process will result, with the exception of the endpoints for the interval, that the functional value at each odd subscripted partition point will be multiplied by 4 and the functional value for each even subscripted partition point will be multiplied by 2. That is:

$$\int_a^b f(x)\,dx = \frac{\Delta x}{3}\left(f\left(x_0\right) + 4f\left(x_1\right) + 2f\left(x_2\right) + 4f\left(x_3\right) + \ldots + 4f\left(x_{n-1}\right) + f\left(x_n\right)\right)$$

Example 13: Find the area under the curve $y = e^{-x^2}$ on the interval $[-1, 2]$ using 6 partitions.

Solution: The graph of the function is:

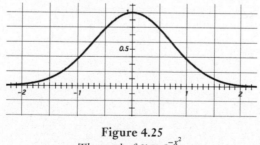

Figure 4.25
The graph of $y = e^{-x^2}$.

The width of each interval is 0.5 and the partition endpoints are $[-1, -0.5]$, $[-0.5, 0]$, $[0, 0.5]$, $[0.5, 1]$, $[1, 1.5]$, and $[1.5, 2]$. Drawing arcs over the partitions shows:

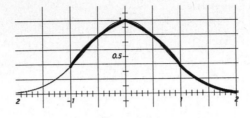

Figure 4.26

Simpson's Rule is applied to $y = e^{-x^2}$ *on the interval* $[-1, 2]$.

The estimate for the integral $\int_{-1}^{2} e^{-x^2}\,dx$ is:

$$\frac{0.5}{3}\left(f(-1) + 4f(-0.5) + f(0)\right) + \left(f(0) + 4f(0.5) + f(1)\right) + \left(f(1) + 4f(1.5) + f(2)\right) = 1.62899$$

⬦	A in	B out	C simpson	D
=	=seq(x,x,-	=q(in)		
1	-1	e^-1	0.367879	f(x0)
2	-0.5	0.778801	3.1152	4*f(x1)
3	0.	1.	2.	2*f(x2)
4	0.5	0.778801	3.1152	4*f(x3)
5	1.	0.367879	0.735759	2*f(4)
6	1.5	0.105399	0.421597	4*f(x5)
7	2.	0.018316	0.018316	f(x6)

Figure 4.27

The spreadsheet display from the TI-Nspire shows the coordinate of the partition and the functional value at that point.

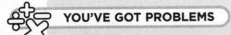

YOU'VE GOT PROBLEMS

Problem 5: Use Simpson's Rule with 6 partitions to evaluate $\int_{0}^{3} \sqrt{x^3 + 4}\,dx$.

The Least You Need to Know

- Areas under a curve can be approximated with rectangles, trapezoids, and parabolic arcs.

- The definite integral represents the area under the curve within the stated bounds of integration.

- The area between two curves is computed with the difference of the functions on the stated interval.

- Technology is very useful in three of the preceding statements.

CHAPTER 5

Volumes and Areas of Solids of Revolutions

We've dealt with a review of Calculus I and what happens in the *xy*-plane—limits, rates of change, and accumulation. Now we get to look at a few three-dimensional issues (the operative word being *few*). What happens when the region formed in a two-dimensional plane is rotated around a line in that plane? The result is a three-dimensional solid. In this chapter, we will explore the process of computing the volume and surface area for these solids.

In This Chapter

- Calculating volumes of solids using disks, washers, and shells
- Determining the distance from point *A* to point *B*
- Understanding areas of solids

The basic concept about how volume is computed is similar to a winter ice storm. It begins with the slimmest sheen of ice laying on the surface. As the precipitation continues to fall, the layer of ice grows thicker and thicker until it is easily seen (and the wise drivers do so from the safety of their homes rather than their cars). The technique translates into calculus as …

1. Identifying an area: A.

2. Multiplying that area by the slimmest of thickness: Δx.

3. Doing so allows accumulation to occur: $\sum A\Delta x$.

4. Applying the notion of the limit means $\sum A\Delta x$ becomes $\int A\,dx$, and a formula is born!

Volumes of Solids with Defined Cross Sections

Let's look at a couple sketches.

Example 1:

Sketch the graph of the parabola $f(x) = 4 - x^2$ on a piece of graph paper for the interval $[-2, 2]$.

1. Connect the points $(-1,0)$ and $(-1, 3)$ with a line segment.

2. Connect $(0,0)$ to $(0,4)$ and $(1,0)$ to $(1,3)$—also with line segments.

3. Use each line segment to construct an equilateral triangle perpendicular to the plane containing the graph paper. That is, imagine the graph paper is flat on a table and the triangles are growing out of the table.

4. Now imagine these triangles being drawn for every pair of points $(b,0)$ and $(b, f(b))$ on the interval $[-2, 2]$. You would have an interesting looking solid.

5. What is the volume of this solid?

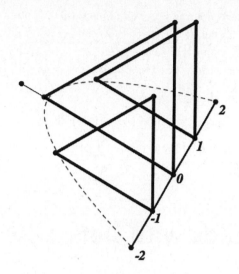

Figure 5.1

A solid with a parabolic base with cross sections perpendicular to the x-axis are equilateral triangles.

Example 2:

Now, sketch the graph of the parabola $f(x) = 4 - x^2$ on the interval $[-2, 2]$ on another piece of graph paper.

1. Connect the pair of points $(-2, 0)$ and $(2, 0)$ with a line segment.

2. Repeat this for the points $(-1, 3)$, and $(1,3)$.

3. Again, construct equilateral triangles along each segment perpendicular to the plane containing the parabola. You have another interesting looking solid but not the same as the solid constructed in the previous diagram.

4. What is the volume of this solid?

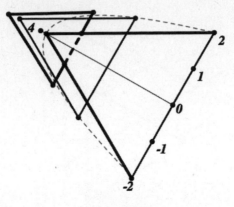

Figure 5.2

A solid with a parabolic base whose cross sections perpendicular to the y-axis are equilateral triangles.

CRITICAL POINT

Four area formulas from geometry are frequently used with this topic. The area of a square with side of length s is s^2, the area of an equilateral triangle with sides of length s is $\frac{s^2\sqrt{3}}{4}$, the area of an isosceles right triangle with a hypotenuse length of s is $\frac{s^2}{4}$, and the area of the semicircle with a diameter s is $\frac{s^2\pi}{8}$.

In the case of the solid shown in earlier Example 1, the volume of the solid formed is $\int A\,dx$ becomes $\int_{-2}^{2} \frac{\left(4-x^2\right)^2 \sqrt{3}}{4}\,dx$. Simplify the integral to $\frac{\sqrt{3}}{4}\int_{-2}^{2}\left(4-x^2\right)^2 dx = \frac{\sqrt{3}}{4}\int_{-2}^{2} 16 - 8x^2 + x^4\,dx$. This becomes $\frac{\sqrt{3}}{4}\left(16x - \frac{8}{3}x^3 + \frac{1}{5}x^5\right)\Big|_{-2}^{2} = \frac{128\sqrt{3}}{15}$.

The case for the solid in Example 2 is more complicated:

1. The lengths of the sides of the triangles are measured by the difference of their x-coordinates. Consequently, the problem needs to be rewritten with y as the independent variable rather than x.

2. If $y = 4 - x^2$, then $y - 4 = -x^2$ so that $x = \pm\sqrt{4 - y}$ with the bounds of integration being $y = 0$ to $y = 4$.

3. The graph of $x = \pm\sqrt{4 - y}$ will generate the right hand side of the parabola.

4. Because of the symmetry of the parabola, the length of each side of the equilateral triangle will be twice this value.

5. The volume of the solid is $\frac{\sqrt{3}}{4}\int_0^4 \left(2\sqrt{4 - y}\right)^2 dy = \sqrt{3}\int_0^4 4 - y\,dy = \sqrt{3}\left(4y - \frac{1}{2}y^2\right)\Big|_0^4 = 8\sqrt{3}$.

Rather than have the cross sections be equilateral triangles, what if they were semicircles? Each segment from $(b,0)$ to $(b,f(b))$ would be the diameter of the semicircle.

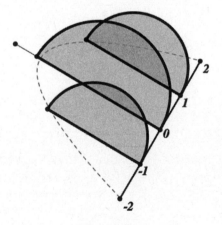

Figure 5.3

A solid with a parabolic base with cross sections perpendicular to the x-axis that are semicircles.

The volume for this solid is $\frac{\pi}{8}\int_{-2}^{2}\left(4-x^2\right)^2 dx = \frac{64\pi}{15}$.

If the cross sections are isosceles right triangles with the hypotenuse in the plane of the parabola, the volume of the solid is $\frac{1}{4}\int_{-2}^{2}\left(4-x^2\right)^2 dx = \frac{128}{15}$.

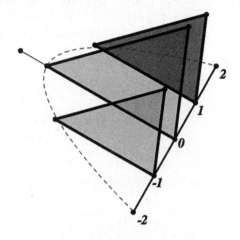

Figure 5.4

A solid with a parabolic base with cross sections perpendicular to the x-axis that are isosceles right triangles.

YOU'VE GOT PROBLEMS

Problem 1: Let R be the region bounded by the graph of $f(x) = \sqrt{x+1}$, the x-axis, $x = 0$, and $x = 4$. Find the volume of the region formed with R as the base if the cross sections perpendicular to the plane of R and to the x-axis are:

    (a) Squares

    (b) Semicircles

    (c) Equilateral triangles

    (d) Isosceles right triangles with the hypotenuse in the plane of R

Disks and Washers

Think of a sketch showing a horizontal line and a segment drawn perpendicular to that line. Rotate the perpendicular line segment around the horizontal line. "Turn" the sketch so that you are looking down the line at the sketch. If one endpoint of the line segment was on the line, you will see a disk (a circle and its interior). If both endpoints of the segment are off the line and on the same side of the line, the figure created will be a washer (concentric circles with the region between them shaded but the interior of the smaller circle not shaded). This is the basis for computing the area, A, needed to compute volume.

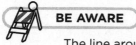 **BE AWARE**

The line around which a region is rotated should not pass through the interior of the region. This creates a problem that we do not want to deal with at this point in our study.

Example 3: The triangular region determined by the x-axis, the y-axis, and the line $2x + y = 4$ is rotated about the x-axis. What is the volume of the solid formed?

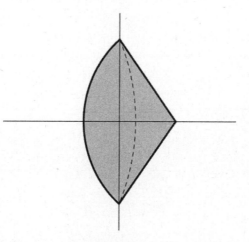

Figure 5.5
The cone formed when the line $2x + y = 4$ is rotated about the x-axis.

Solution:

1. If you look at a cross section from the perspective of the "end" of the x-axis, you would see this:

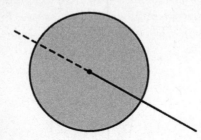

Figure 5.6

A cross section of the cone formed when the line $2x + y = 4$ is rotated about the x-axis is a disk.

2. Each cross section is a circle with radius $y = 4 - 2x$.

3. Therefore, $\int A\,dx$ becomes $\pi \int_0^2 (4 - 2x)^2\, dx = \frac{32\pi}{3}$. (This agrees with the formula for the volume of a cone with base radius = 4 and height 2.)

Example 4: The triangular region determined by the x-axis, the y-axis, and the line $2x + y = 4$ is rotated about the y-axis. What is the volume of the solid formed?

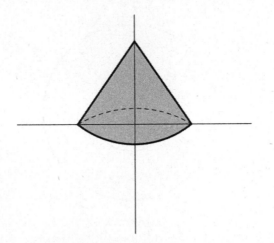

Figure 5.7

The cone formed when the line $2x + y = 4$ is rotated about the y-axis.

Solution: Each cross section is a circle (as seen from the "end" of the y-axis) with radius $x = \frac{4-y}{2}$ so the volume of the solid is $\pi \int_0^4 (\frac{4-y}{2})^2\, dy = \frac{16\pi}{3}$.

Example 5: The triangular region determined by the x-axis, the y-axis, and the line $2x + y = 4$ is rotated about the line $y = -2$. What is the volume of the solid formed?

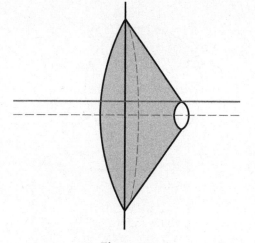

Figure 5.8
The solid formed when the line $2x + y = 4$ is rotated about the line $y = -2$.

Solution:

1. If you look at a cross section from the perspective of the "end" of the x-axis, you would see this:

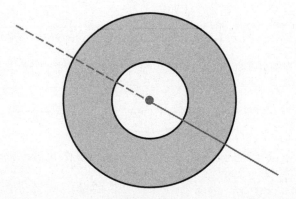

Figure 5.9
A cross section of the solid formed when the line $2x + y = 4$ is rotated about the line $y = -2$
is a washer.

2. The radius of the smaller circle is 2 (the distance from $y = -2$ to the x-axis).

3. The distance of the larger radius is $4 - 2x + 2$ (the distance from $y = -2$ to the line $y = 4 - 2x$), or $y = 6 - 2x$.

4. The area of a cross section is computed as $\pi((6 - 2x)^2 - 2^2)$ so the volume of the solid is
$$\pi \int_0^2 (6 - 2x)^2 - 4\, dx = \frac{80\pi}{3}.$$

Example 6: The triangular region determined by the x-axis, the y-axis, and the line $2x + y = 4$ is rotated about the line $x = 4$. What is the volume of the solid formed?

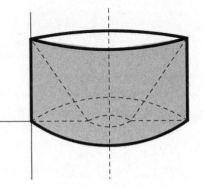

Figure 5.10
The solid formed when the line $2x + y = 4$ is rotated about the line $x = 4$.

Solution:

1. The figure is that of a cylinder with a truncated cone removed from its middle.

2. If you look at a cross section from the perspective of the "end" of the line $x = 4$, you would see a washer.

3. The radius of the smaller circle is $4 - \frac{4-y}{2} = \frac{4+y}{2}$ (the distance from $x = 4$ to the line $x = \frac{4-y}{2}$).

4. The distance of the larger radius is 4 (the distance from the y-axis to the line $x = 4$).

5. The area of a cross section is $\pi\left(4^2 - \left(\frac{4+y}{2}\right)^2\right)$ so the volume of the solid is
$$\pi \int_0^4 \left(16 - \left(\frac{4+y}{2}\right)^2\right) dy = \frac{80\pi}{3}.$$

YOU'VE GOT PROBLEMS

Problem 2: Let R be the region bounded by the axes and the line $4x + 3y = 12$. Find the volume of the solid formed when R is rotated around the line $y = -2$.

Example 7: The region bounded by the graphs of $y = x^2$ and $y = x + 2$ is rotated around the x-axis. Find the volume of the solid formed.

Solution:

1. A graph of the two functions shows that they intersect at the points $(-1,1)$ and $(2,4)$.

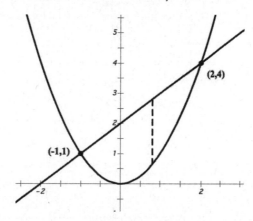

Figure 5.11

A segment is drawn perpendicular to the x-axis in the region bounded by the graphs of $y = x + 2$ and $y = x^2$.

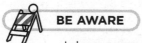 **BE AWARE**

I drew a segment at an arbitrary value of x in the interval $[-1, 2]$. I use this figure to determine if the cross section of the solid is a disk or a washer.

2. As you can see, there is a gap between the lower endpoint of the segment and the x-axis so the cross section will be a washer.

3. The larger radius of the washer is the distance from the x-axis to the line, $x + 2$.

4. The smaller radius is the distance from the x-axis to the parabola, x^2.

5. Therefore, the area of the cross section is $\pi\left((x+2)^2 - \left(x^2\right)^2\right)$.

6. This means that the volume of the solid is $\pi\int_{-1}^{2}\left((x+2)^2 - \left(x^2\right)^2\right)dx = \frac{72\pi}{5}$.

Example 8: The region bounded by the x-axis and $f(x) = \sin(x)$ from $x = 0$ to $x = \pi$ is rotated around the x-axis. Find the volume of the solid formed.

Solution: A cross section of the region will be a disk so the volume is $\pi\int_{0}^{\pi}\sin^2(x)dx =$

$\pi\left(\frac{1}{2}x - \frac{1}{4}\sin(2x)\right)\Big|_{0}^{\pi} = \pi\left(\frac{1}{2}\pi - \frac{1}{4}\sin(2\pi)\right) - \left(\frac{1}{2}(0) - \frac{1}{4}\sin(0)\right) = \frac{\pi^2}{2}$. (The antiderivative was found in Chapter 3's Example 7.)

Example 9: The region bounded by the x-axis and $f(x) = \cos(x)$ from $x = 0$ to $x = \frac{\pi}{2}$ is rotated around the line $y = -1$. Find the volume of the solid formed.

Solution:

1. A cross section for this solid will be a washer with larger radius $\cos(x) + 1$ and smaller radius 1.

2. The volume of the solid formed is $\pi\int_{0}^{\pi/2}\left(\cos(x)+1\right)^2 - 1\,dx = \pi\int_{0}^{\pi/2}\cos^2(x) + 2\cos(x)\,dx$.

3. Use the trigonometric identity $\cos(2x) = 2\cos^2(x) - 1$, or $\cos^2(x) = \frac{1}{2}\cos(2x) + \frac{1}{2}$ to change the integral to $\pi\int_{0}^{\pi/2}\frac{1}{2}\cos(2x) + \frac{1}{2} + 2\cos(x)\,dx = \pi\left(\frac{1}{4}\sin(2x) + \frac{x}{2} + 2\sin(x)\right)\Big|_{0}^{\pi/2} =$

$\pi\left[\left(\frac{1}{4}\sin(\pi) + \frac{\pi}{4} + 2\sin\left(\frac{\pi}{2}\right)\right) - \left(\frac{1}{4}\sin(0) + 0 + 2\sin(0)\right)\right] = \pi\left(\frac{\pi}{4} + 2\right) = \frac{\pi^2 + 8\pi}{4}$.

Example 10: The graph of $y = 4\cos(2x)$ intersects the x-axis at point K, as shown in the following figure.

C is the region bounded by the graph of $y = 4\cos(2x)$ and the line segment joining the y-intercept of this graph to point K. Find the volume of the solid formed when C is rotated around the x-axis.

Solution:

1. Point K is the first positive value of x when $\cos(2x) = 0$.

2. You know that $\cos(A) = 0$ when $A = \frac{\pi}{2}$, so $2x = \frac{\pi}{2}$ implies that $x = \frac{\pi}{4}$.

3. The line joining $(0,4)$ (the y-intercept) to $\left(\frac{\pi}{4},0\right)$ has slope $\frac{0-4}{\frac{\pi}{4}-0} = \frac{-16}{\pi}$, so the equation of the line is $y = \frac{-16}{\pi}x + 4$.

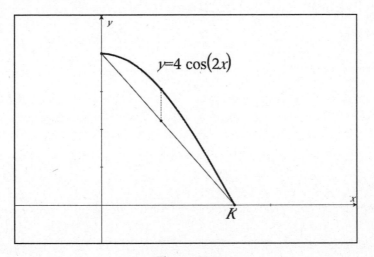

Figure 5.12
A line segment is drawn between the intercepts of the graph of $y = 4\cos(2x)$.

4. Using the arbitrarily drawn segment within the bounds of C, you can see that a cross section of the solid will be a washer.

5. The area of this cross section is $\pi\left(\left(4\cos(2x)\right)^2 - \left(\frac{-16}{\pi}x + 4\right)^2\right)$.

6. This is a particularly ugly Area function so I chose to use the integration feature on my graphing calculator to determine that $\pi\int_0^{\pi/4}\left(\left(4\cos(2x)\right)^2 - \left(\frac{-16}{\pi}x + 4\right)^2\right)dx = \frac{2\pi^2}{3}$.

Example 11: Let Q be the region bounded by the f(x) = x^3, $y = 8$, and the y-axis. Find the volume of the solid formed when Q is revolved around the line $y = 8$ and the line $x = 2$.

Solution: The graph of the Q is:

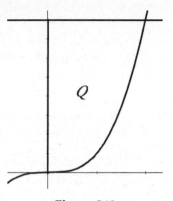

Figure 5.13
Q is the region formed by the y-axis, y = 8, and y = x^3.

1. The cross section of the solid when Q is rotated around the line $y = 8$ will be a disk with radius $8 - x^3$. (Draw a segment at an arbitrary point on the graph of $y = x^3$ to the line $y = 8$. The length of this segment is $8 - x^3$.) The bounds of integration are from $x = 0$ to $x = 2$. Therefore, the volume of the solid is $\pi \int_0^2 (8 - x^3)^2 \, dx = \pi \int_0^2 64 - 16x^3 + x^6 \, dx = \pi \left(64x - 4x^4 + \frac{1}{7}x^7 \Big|_0^2 \right) = \frac{576\pi}{7}$.

2. The cross section of the solid when Q is rotated around the line $x = 2$ will be a washer.

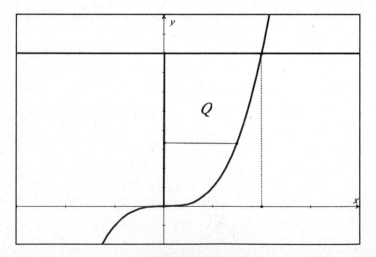

Figure 5.14
Region Q with an arbitrarily selected segment drawn perpendicular to the axis of rotation.

This solid will be similar to that of Example 6. There is a cylinder of radius 2 with a shape removed from its center. The radii of the washer will be 2 and $2 - \sqrt[3]{y}$. The bounds of integration will be from $y = 0$ to $y = 8$.

BE AWARE

Because the line of rotation is vertical, the equation needs to be rewritten as $x = f(y)$ rather than $y = f(x)$. In this case, $y = x^3$ becomes $x = \sqrt[3]{y}$.

The volume of the solid formed when Q is rotated around the line $x = 2$ is $\pi \int_0^8 2^2 - \left(2 - \sqrt[3]{y}\right)^2 dy$ $= \frac{144\pi}{5}$.

Example 12: Let P be the region bounded by the graphs of $y = e^{-x}$ and $y = 2 - x$. What is the volume of the solid formed when P is rotated about the x-axis?

Solution: Using the calculator to sketch the graphs of the two functions shows that they intersect at two points. Using a calculator makes finding the points of intersection much easier than it is to do so algebraically.

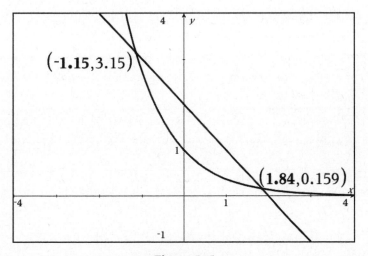

Figure 5.15
The graph shows the region formed by the intersection of $y = e - x$ and $y = 2 - x$.

1. Store each of the x-coordinates of the points of intersection into variables on your calculator, letting $a = -1.15$ and $b = 1.84$. Hopefully you do not need to physically draw a segment in the region P to determine that the cross section of the solid will be a washer.

2. The larger radius of the washer will be $2 - x$, while the smaller radius will be e^{-x}.

3. The volume of the solid is $\pi \int_a^b (2 - x)^2 - e^{-2x} \, dx = 17.0993$.

YOU'VE GOT PROBLEMS

Problem 3: Find the volume of the region bounded by the x-axis, $y = \dfrac{1}{\sqrt{1+x^2}}$, $x = 0$, and $x = 1$ is revolved around the x-axis.

Cylindrical Shell Method

When a vertical line segment is rotated about a vertical line, the solid formed is a shell (empty cylinder) rather than a disk. The same thing happens when a horizontal segment is rotated around a horizontal line. As discussed earlier in this chapter (see "Disks and Washers"), if infinitely thin layers of the cylinder are laid one upon the other, a thickness develops and a solid is created. The area for the lateral side of a cylinder is $2\pi rh$.

CRITICAL POINT

Using a rectangular sheet of paper with the area l × w, roll the paper along the longer side into a cylinder. The width of the paper becomes the height of the cylinder. The length of the paper becomes the circumference of the circle. So l × w become h × 2 × π × r.

The term $\sum A\Delta x$ becomes $\sum 2\pi rh\Delta x$ and $\int A\,dx$ becomes $\int 2\pi rh\,dx = 2\pi \int rh\,dx$. The task before us as we try to use this material is to determine expressions for the radius and circumference of a cross-sectional cylinder.

Example 13: Use the shell method to determine the volume of the solid formed when the region bounded by $y = 4 - 2x$, $x = 0$, and $y = 0$ is rotated around the line $x = 0$.

Solution:

1. If this problem looks familiar, it's because it's the same problem as Example 4.

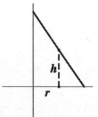

Figure 5.16
The cone formed when the line $2x + y = 4$ is rotated about the y-axis. An arbitrarily selected segment in the region is drawn perpendicular to the x-axis.

2. The radius of the cylinder will be x.

3. The height of the cylinder is $4 - 2x$.

4. The volume of the solid formed is $2\pi \int_0^2 x(4 - 2x)\,dx = 2\pi \int_0^2 4x - 2x^2\,dx = 2\pi \left(2x^2 - \frac{2}{3}x^3 \Big|_0^2 \right) = \frac{16\pi}{3}$.

Example 14: The region bounded by the graphs of $y = x^3$, $y = 8$, and $x = 0$ are rotated around the line $x = 2$. Find the volume of the solid formed.

Solution: This problem is part 2 of Example 11.

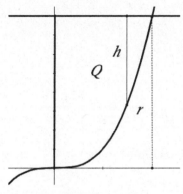

Figure 5.17
Q is the region formed by the y-axis, $x = 8$, and $y = x^3$. An arbitrarily selected segment in the region is drawn perpendicular to the x-axis.

1. The radius of the cylinder is $2 - x$.

2. The height of the cylinder is $8 - x^3$.

3. The volume of the solid is $2\pi \int_0^2 (2 - x)(8 - x^3)\,dx = \frac{144\pi}{5}$.

✏️ (**CRITICAL POINT**)

In general, it is usually easier to use the cylindrical shell method when the axis of rotation is a vertical line and to use the disk or washer methods when the axis of revolution is a horizontal line.

Example 15: The region bounded by the graphs of $y = x^2$ and $y = 2x$ is rotated around the y-axis. Find the volume of the solid formed.

Solution:

    1. The two graphs intersect at (0,0) and (2,4).

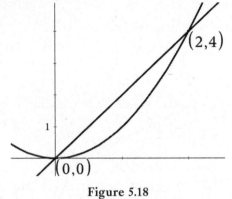

Figure 5.18
The region formed by the intersection of $y = x^2$ and $y = 2x$.

    2. Picking any arbitrary point in the interval [0, 2], the radius of the cylinder is x.

    3. The height of the cylinder is $2x - x^2$.

    4. Therefore, the volume of the solid formed is $2\pi \int_0^2 x(2x - x^2)\,dx = \frac{8\pi}{3}$.

YOU'VE GOT PROBLEMS

Problem 4: The region bounded by the graphs of $y = 4x - x^2$ and $y = 2x$ is rotated about the y-axis. Find the volume of the solid formed.

Example 16: The region bounded by the graphs of $y = x^2$ and $y = 2x$ is rotated around the line $x = 2$. Find the volume of the solid formed.

Solution: Picking any arbitrary point in the interval [0, 2], the radius of the cylinder is $2 - x$ and the height of the cylinder is $2x - x^2$. Therefore, the volume of the solid formed is $2\pi \int_0^2 (2 - x)(2x - x^2)\,dx = \frac{8\pi}{3}$.

Example 17: The region bounded by the graphs of $y = x^2$ and $y = 2x$ is rotated around the line $x = 3$. Find the volume of the solid formed.

Solution: Picking any arbitrary point in the interval $[0, 2]$, the radius of the cylinder is $3 - x$ and the height of the cylinder is $2x - x^2$. Therefore, the volume of the solid formed is $2\pi \int_0^2 (3 - x)(2x - x^2) \, dx = \frac{16\pi}{3}$.

Example 18: The line drawn tangent to the graph of $f(x) = 9x - x^3$ has a tangent line drawn to it at $x = 2$. The equation of the tangent line is $y = -3x + 16$. Two regions are indicated on the graph. R is bounded by the graph of $f(x)$ and the x-axis. S is bounded by the graph of $f(x)$, the tangent line, and the y-axis.

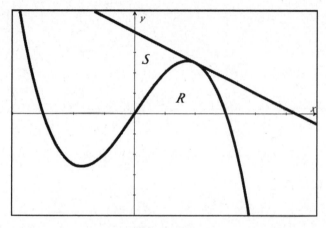

Figure 5.19
Regions formed by the graphs of $f(x) = 9x - x^3$ *and* $y = -3x + 16$.

1. Find the volume of the solid formed when R is rotated around the y-axis.

2. Find the volume formed when S is rotated around the x-axis.

3. Find the volume formed when S is rotated around the line $x = 3$.

Solution:

1. Use cylindrical shells with radius $= x$ and height $= 9x - x^3$. The volume equals

$$2\pi\int_0^3 x(9x - x^3)\,dx = 2\pi\int_0^3 9x^2 - x^4\,dx = 2\pi\left(3x^3 - \tfrac{1}{5}x^5\right)\Big|_0^3 =$$
$$2\pi(0) - 2\pi\left(\tfrac{-243}{5} + 81\right) = \tfrac{324\pi}{5}.$$

2. Use disks to compute this volume with the larger circle having radius $-3x + 16$ and the smaller circle having radius $9x - x^3$. The volume is $\pi\int_0^2 (-3x + 16)^2 - \left(9x - x^3\right)^2\,dx =$

$$\pi\int_0^2 9x^2 - 96x + 256 - \left(81x^2 - 18x^4 + x^6\right)dx.$$ Combine like terms to get

$$\pi\int_0^2 -x^6 + 18x^4 - 72x^2 - 96x + 256\,dx = \pi\left(\tfrac{-1}{7}x^7 + \tfrac{18}{5}x^5 - 24x^3 - 48x^2 + 256x\right)\Big|_0^2 = \tfrac{7872\pi}{35}$$

(or 706.589).

3. Use cylindrical shells with radius $= 3 - x$ and height $= (-3x + 16) - (9x - x^3)$
   $= x^3 - 12x + 16$. The volume of the solid is $2\pi\int_0^2 (3 - x)\left(x^3 - 12x + 16\right)dx =$

$$2\pi\int_0^2 3x^3 - 36x + 48 - x^4 + 12x^2 - 16x\,dx = 2\pi\int_0^2 -x^4 + 3x^3 + 12x^2 - 52x + 48\,dx =$$

$$2\pi\left(\tfrac{-1}{5}x^5 + \tfrac{3}{4}x^4 + 4x^3 - 26x^2 + 48x\right)\Big|_0^2 = 2\pi\left(\tfrac{-1}{5}(32) + \tfrac{3}{4}(16) + 4(8) - 26(4) + 48(2)\right) =$$

$\tfrac{296\pi}{5}$ (or 185.982).

Arc Length

You learned in geometry that the distance from point $A(x_1, y_1)$ to $B(x_2, y_2)$ is

$\sqrt{\left(x_2 - x_1\right)^2 + \left(y_2 - y_1\right)^2}$. How do we measure the distance from point A to point B if the path taken follows the graph of a function?

Earlier in the book, I suggested that the essence of calculus is putting algebra under a microscope.

1. If we look at a graph of $y = f(x)$ and at two points on the curve (x,y) and $(x + \Delta x, y + \Delta y)$:

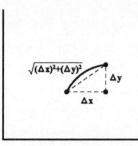

Figure 5.20
Use the distance formula between the points (x, y) and $(x + \Delta x, y + \Delta y)$ to estimate the length of the arc between two points on a curve.

2. The distance between these two points is $\sqrt{(x + \Delta x - x)^2 + (y + \Delta y - y)^2}$ =
$\sqrt{(\Delta x)^2 + (\Delta y)^2}$ = $\Delta x \sqrt{1 + \left(\frac{\Delta y}{\Delta x}\right)^2}$.

3. If we add up the various pieces of segments that approximate the curve, the distance from A to B is approximately $\sum \sqrt{1 + \left(\frac{\Delta y_i}{\Delta x_i}\right)^2} \, \Delta x$.

4. As we saw with the argument for area under a curve and the volume of solids of revolution, if we let Δx get very small (in math language, $\lim\limits_{\Delta x \to 0}$), $\sum \sqrt{1 + \left(\frac{\Delta y_i}{\Delta x_i}\right)^2} \, \Delta x$ becomes $\int \sqrt{1 + \left(\frac{dy}{dx}\right)^2} \, dx$.

5. Clearly, this is the same as $\int \sqrt{1 + (f'(x))^2} \, dx$.

6. If x is a function of y, the formula becomes $\int \sqrt{1 + \left(\frac{dx}{dy}\right)^2} \, dy$.

Example 19: Determine the length of the arc along the graph of $f(x) = \sqrt{1-x^2}$ on the interval $[0, 1]$.

Solution:

1. $f'(x) = \frac{-x}{\sqrt{1-x^2}}$, so $(f'(x))^2 = \frac{x^2}{1-x^2}$. The length of the arc is $\int_0^1 \sqrt{1+\frac{x^2}{1-x^2}}\,dx =$

   $\int_0^1 \sqrt{\frac{1-x^2+x^2}{1-x^2}}\,dx = \int_0^1 \sqrt{\frac{1}{1-x^2}}\,dx = \int_0^1 \frac{1}{\sqrt{1-x^2}}\,dx = \sin^{-1}(x)\Big|_0^1 = \sin^{-1}(1) - \sin^{-1}(0) = \frac{\pi}{2}$.

2. We have just shown that the distance along one quarter of the circumference of the unit circle is $\frac{\pi}{2}$.

3. Therefore, the circumference of the entire unit circle is four times this amount or 2π.

Surprised? Probably not. But I hope you do appreciate this is likely the first time anyone has proven to you that the circumference formula is accurate.

Example 20: Find the length of the curve of the function $f(x) = \ln|\sec(x)|$ on the interval $\left[0, \frac{\pi}{3}\right]$.

Solution:

1. The derivative for $f(x)$ is $f'(x) = \frac{\sec(x)\tan(x)}{\sec(x)} = \tan(x)$.

2. Therefore, the length of the curve is $\int_0^{\pi/3} \sqrt{1+\tan^2(x)}\,dx$.

3. Use the Pythagorean identity $1 + \tan^2(x) = \sec^2(x)$ to rewrite the integral as

   $\int_0^{\pi/3} \sqrt{\sec^2(x)}\,dx = \int_0^{\pi/3} |\sec(x)|\,dx$.

4. On the interval $\left[0, \frac{\pi}{3}\right]$, $\sec(x) > 0$, so the integral can be written as $\int_0^{\pi/3} \sec(x)\,dx$.

5. Recall from Example 12 of Chapter 3 that $\int \sec(x)\,dx = \ln|\sec(x) + \tan(x)|$ so

   $\int_0^{\pi/3} \sec(x)\,dx = \ln|\sec(x) + \tan(x)|\Big|_0^{\pi/3} = \ln|2 + \sqrt{3}\,| - \ln|1 + 0| = \ln|2 + \sqrt{3}\,|$.

Example 21: Find the length of the arc along the parabola $y^2 = 8x$ on the interval $[1,4]$.

Solution:

1. Use implicit differentiation to determine that $2y\frac{dy}{dx} = 8$ so that $\frac{dy}{dx} = \frac{4}{y}$ and

   $\left(\frac{dy}{dx}\right)^2 = \frac{16}{y^2} = \frac{16}{8x} = \frac{2}{x}$.

2. Therefore, the length of the arc along the parabola is $\int_1^4 \sqrt{1+\frac{2}{x}}\,dx = \int_1^4 \sqrt{\frac{x+2}{x}}\,dx$.

3. This does not fit any of the integration formulas that we know at this time, so let's use the graphing calculator to get the answer. $\int_1^4 \sqrt{\frac{x+2}{x}}\, dx$ = 4.1424. (My calculator gave the answer $2\ln\left(\frac{(\sqrt{6}+2)(\sqrt{3}-1)}{2}\right)+\left(2\sqrt{2}-1\right)(\sqrt{3})$. We'll have to look at this problem again after we learn some new integration techniques in Chapter 7.)

YOU'VE GOT PROBLEMS

Problem 5: The graph of $\frac{x^2}{25}+\frac{y^2}{16}=1$ is an ellipse. Find the length of the arc in the first quadrant on the interval [0,5].

Surface Area

The process for computing the surface area of a solid of revolution is similar to how we computed volumes using the shell method. We said that the area of a cylinder is $2\pi rh$. (See "Cylindrical Shell Method," earlier in this chapter.) As we looked to compute volume, the values for r and h were dependent upon the function in question and the line around which the region was being rotated. The formula for the surface area of a solid of revolution when a curve is rotated about the x-axis is $2\pi \int f(x)\sqrt{1+(f'(x))^2}\, dx$.

Example 22: Find the surface area of the solid formed when the section of the graph of f(x) = x^3 from x = 0 to x = 4 is rotated about the x-axis.

Solution:

1. The surface area is $2\pi \int_0^4 x^3 \sqrt{1+(3x^2)^2}\, dx = 2\pi \int_0^4 x^3 \sqrt{1+9x^4}\, dx$.

2. Let $u = 1 + 9x^4$, then $du = 36x^3 dx$ so that $\frac{1}{36}du = x^3 dx$.

3. Transforming the bounds of integration, when x = 0, u = 1, and when x = 4, u = 2305.

4. The transformed integral is $\dfrac{2\pi}{36}\int_1^{2305}\sqrt{u}\, du = \frac{\pi}{18}\cdot\frac{2}{3}u^{3/2}\Big|_1^{2305} = \frac{\pi\left(2305\sqrt{2305}-1\right)}{27}$, or approximately 12876.226.

Example 23: Find the surface area of the solid formed when a section of the graph $x^2 - y^2 = 25$ from $y = 0$ to $x = 10$ is rotated around the x-axis.

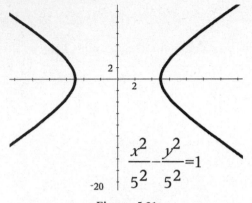

Figure 5.21
A graph of the hyperbola $x^2 - y^2 = 25$.

Solution:

1. When $y = 0$, $x = 5$ (this is where the hyperbola crosses the x-axis). We'll work with the portion of the hyperbola in the first quadrant (because we can—we could just as easily work with the portion of the hyperbola in the fourth quadrant but that would require us to be careful with a bunch of negative signs).

2. Rewrite the equation to be $y^2 = x^2 - 25$ so that $y = \sqrt{x^2 - 25}$, $\frac{dy}{dx} = \frac{x}{\sqrt{x^2 - 25}}$, and $\left(\frac{dy}{dx}\right)^2 = \frac{x^2}{x^2 - 25}$.

3. The surface area of the solid formed is $2\pi \int_5^{10} \left(\sqrt{x^2 - 25}\right)\left(\sqrt{1 + \frac{x^2}{x^2 - 25}}\right) dx =$

   $2\pi \int_5^{10} \left(\sqrt{x^2 - 25}\right)\left(\sqrt{\frac{x^2 - 25 + x^2}{x^2 - 25}}\right) dx = 2\pi \int_5^{10} \sqrt{2x^2 - 25}\, dx$.

4. Use the graphing calculator to determine the surface area is 291.588.

Example 24: Find the surface area of the solid formed when a section of the graph $x^2 - y^2 = 25$ from $y = 0$ to $x = 10$ is rotated around the y-axis.

Solution:

1. When $x = 10$, $y = 5\sqrt{3}$.

2. This time we'll rewrite the equation to be $x^2 = y^2 + 25$ so that $x = \sqrt{y^2 + 25}$, $\frac{dx}{dy} = \frac{y}{\sqrt{y^2 + 25}}$, and $\left(\frac{dx}{dy}\right)^2 = \frac{y^2}{y^2 + 25}$.

3. The surface area of the solid formed is $2\pi\int_{0}^{5\sqrt{3}}\left(\sqrt{y^2+25}\right)\left(\sqrt{1+\frac{y^2}{y^2+25}}\right)dy =$

$2\pi\int_{0}^{5\sqrt{3}}\left(\sqrt{y^2+25}\right)\left(\sqrt{\frac{y^2+25+y^2}{y^2+25}}\right)dy = 2\pi\int_{0}^{5\sqrt{3}}\left(\sqrt{2y^2+25}\right)dy$.

4. Use the graphing calculator to determine that the surface area is 450.344.

YOU'VE GOT PROBLEMS

Problem 6: The section of the parabola $y = 4x - x^2$ from $x = 0$ to $x = 4$ is rotated around the x-axis. Find the surface area of the solid formed.

The Least You Need to Know

- Compute volumes of solids formed when cross sections to a region are squares, semicircles, equilateral triangles, or isosceles right triangles with the hypotenuse in the plane of the region by applying area formulas from plane geometry.

- Compute volumes for solids of revolution, when the cross section is a disk or a washer, using the formula for the area of a circle.

- Compute volumes for solids of revolution with the cylindrical shell method using the formula for the surface area of a cylinder.

- Compute the length of an arc formed along the graph of a function by applying the Pythagorean Theorem.

- Compute the surface area for a solid of revolution by applying the formula for computing arc length.

PART 3

More Definite and Indefinite Integrals

You saw in Chapter 5 that there are a few integrals we do not yet know how to evaluate. We continue to explore integration in this part of the book.

You learn that certain combinations of polynomial and transcendental functions (those functions that cannot be written as polynomials—trigonometric functions, exponential functions, and logarithmic functions in particular) can be worked with using integration by parts.

We also extend our study of trigonometric functions, rational functions, and irrational functions.

When you are done with this part, go to Appendix B, where there is a variety of integration problems for you to try. The goal is for you to be able to distinguish one type of problem from another.

CHAPTER 6

More Integration Techniques

If you read Chapters 2 and 3, you know how to integrate $\int x^n \, dx$ for all real values of x. You know that the problem $\int x\sqrt{x} \, dx$ is the same as $\int x^{3/2} \, dx$, but what do you do for the problem $\int x\sqrt{x+1} \, dx$? You know how to deal with $\int x\sqrt{1+x^2} \, dx$ but not $\int \sqrt{1+x^2} \, dx$.

Yes, it is true that we could let the graphing calculator get the answer for us, but where is the fun in that? Our goal for the next three chapters is to extend the integration patterns we recognize and can use. First up is integration by parts.

In This Chapter

* Applying integration by parts
* Understanding polynomials and transcendentals
* Working with two transcendentals

Integration by Parts

You know the product rule for differentiation of $y = f(x)g(x)$ is $y' = f'(x)g(x) + f(x)g'(x)$. If we use differentials to write this property, we get $d(uv) = v\,du + u\,dv$. If we integrate both sides of this equation, we get $\int d(uv) = \int v\,du + \int u\,dv$. The expression $\int d(uv) = uv$ so the equation is now $uv = \int v\,du + \int u\,dv$. Alter this equation to get $\int u\,dv = uv - \int v\,du$. This is the *formula for integration by parts*.

> **DEFINITION**
>
> The **formula for integration by parts** is $\int u\,dv = uv - \int v\,du$.

Example 1: Evaluate $\int x\sqrt{x+1}\,dx$.

Solution:

1. The key issue in this problem is to determine which part of the integrand is u and which part is dv.

2. In problems involving polynomials, let the polynomial be u.

3. What you will see happen is that the polynomial can be eliminated from the integrand, allowing for a problem that you can integrate.

4. I find that it is usually easier to write the pieces of the integration by parts formula in columns rather than all on one line:

$$u = x \qquad\qquad dv = \sqrt{x+1}\,dx$$

$$du = dx \qquad\qquad v = \tfrac{2}{3}(x+1)^{3/2}$$

5. Notice that $u\,dv$ is the first row.

6. The rest of the problem is now to take the main diagonal (upper left to lower right) to get uv and subtract from it the second row to get $v\,du$:

$$\int x\sqrt{x+1}\,dx = \tfrac{2}{3}x(x+1)^{3/2} - \int \tfrac{2}{3}(x+1)^{3/2}\,dx$$

$$\int x\sqrt{x+1}\,dx = \tfrac{2}{3}x(x+1)^{3/2} - \tfrac{2}{3}\left(\tfrac{2}{5}\right)(x+1)^{5/2} + C = \tfrac{2}{3}x(x+1)^{3/2} - \tfrac{4}{15}(x+1)^{5/2} + C$$

7. Take the derivative of this final answer to verify that it is correct.

There are times when one application of integration by parts will not be enough to get the final answer.

YOU'VE GOT PROBLEMS

Problem 1: Evaluate $\int 36x^3\sqrt{9x+1}\,dx$.

Example 2: Evaluate $\int x^2\sqrt{x+1}\,dx$.

Solution:

1. Break down the original integrand and examine the four pieces of the formula:

    $u = x^2$ $\qquad\qquad$ $dv = \sqrt{x+1}\,dx$

    $du = 2x\,dx$ $\qquad\qquad$ $v = \frac{2}{3}x^2(x+1)^{3/2}$

2. Apply the formula: $\int x^2\sqrt{x+1}\,dx = \frac{2}{3}x^2(x+1)^{3/2} - \frac{2}{3}\int(x+1)^{3/2}(2x)\,dx =$

    $\frac{2}{3}x^2(x+1)^{3/2} - \frac{4}{3}\int x(x+1)^{3/2}\,dx$.

3. As you can see, we are left with another integral that we do not recognize.

4. However, notice that another application of integration by parts will remove the lone factor x and will get us to the answer.

5. Repeat the process:

    $u = x$ $\qquad\qquad$ $dv = (x+1)^{3/2}\,dx$

    $du = dx$ $\qquad\qquad$ $v = \frac{2}{5}x(x+1)^{5/2}$

BE AWARE

When applying the technique of integration by parts more than once, keep in mind that the integral being replaced is being subtracted from the previous term. This means a negative will be distributed through the replacement values.

$\int x^2\sqrt{x+1}\,dx = \frac{2}{3}x^2(x+1)^{3/2} - \frac{4}{3}\int x(x+1)^{3/2}\,dx =$

$\frac{2}{3}x^2(x+1)^{3/2} - \frac{4}{3}\left(\frac{2}{5}x(x+1)^{5/2} + \frac{2}{5}\int(x+1)^{5/2}\,dx\right)$

$\int x^2\sqrt{x+1}\,dx = \frac{2}{3}x^2(x+1)^{3/2} - \frac{8}{15}x(x+1)^{5/2} + \frac{8}{15}\int(x+1)^{5/2}\,dx =$

$\frac{2}{3}x^2(x+1)^{3/2} - \frac{8}{15}x(x+1)^{5/2} + \frac{8}{15}\left(\frac{2}{7}(x+1)^{7/2}\right) + C$

$\int x^2\sqrt{x+1}\,dx = \frac{2}{3}x^2(x+1)^{3/2} - \frac{8}{15}x(x+1)^{5/2} + \frac{16}{105}(x+1)^{7/2} + C$

There is a graphical approach to doing integration by parts that is called the *Tabular Method*. There are three columns:

- The first column starts with u and will be used to compute the derivatives of u as we work down the column.

- The second column starts with dv and will be used to compute the antiderivatives of dv as we work down the column.

- The third column starts with +1 and alternates between +1 and −1 as we work down the column.

DEFINITION

To complete the **Tabular Method,** work from the first row, first column, moving diagonally down the table. Then start with the second row, first column, moving diagonally down the table. Continue in this manner until you reach a 0 in the first column. If the original problem is an indefinite integral, add the constant of integration, $+ C$, at the end of the problem.

u	dv	+1
x^2	$(x+1)^{1/2}$	+1
$2x$	$\frac{2}{3}(x+1)^{3/2}$	-1
2	$\frac{4}{15}(x+1)^{5/2}$	+1
0	$\frac{8}{105}(x+1)^{7/2}$	-1
		+1

Figure 6.1
The Tabular Method for performing integration by parts on $\int x^2 \sqrt{x+1}\, dx$.

1. The first step results in $\left(x^2\right)\left(\frac{2}{3}(x+1)^{3/2}(1)\right)$.

2. The second step results in $\left(2x\right)\left(\frac{4}{15}(x+1)^{5/2}(-1)\right)$.

3. The third step results in $\left(2\right)\left(\frac{8}{105}(x+1)^{7/2}(1)\right)$.

4. All terms after that will contain a factor of 0.

5. The Tabular Method for integration by parts for the problem $\int x^2\sqrt{x+1}\,dx$ is

$$\left(x^2\right)\left(\tfrac{2}{3}(x+1)^{\frac{3}{2}}\,(1)\right) + \left(2x\right)\left(\tfrac{4}{15}(x+1)^{\frac{5}{2}}\,(-1)\right) + \left(2\right)\left(\tfrac{8}{105}(x+1)^{\frac{7}{2}}\,(1)\right) + C.$$

6. Simplify the terms to get the solution $\tfrac{2}{3}x^2(x+1)^{\frac{3}{2}} - \tfrac{8}{15}x(x+1)^{\frac{5}{2}} + \tfrac{16}{105}(x+1)^{\frac{7}{2}} + C$.

BE AWARE

Once you have decided which functions to designate as u and dv, you must be consistent in this manner if the technique of integration by parts is needed more than once. If you switched which functions were designated u and dv in the second application of the technique, you would return right back to the problem with which you began.

YOU'VE GOT PROBLEMS

Problem 2: Use either method of integration by parts to evaluate $\int x^2\sqrt{8x+3}\,dx$.

Polynomials and Transcendentals

Integration by parts is utilized when the integrand is the product of a polynomial and a transcendental function. Let's begin with a few easy problems and then increase the complexity of the functions.

Example 3: Evaluate $\int x\sin(x)\,dx$.

Solution: Let:

$$u = x \qquad\qquad dv = \sin(x)dx$$

$$du = dx \qquad\qquad v = -\cos(x)$$

$$\int x\sin(x)\,dx = -x\cos(x) - \int -\cos(x)\,dx = -x\cos(x) + \int \cos(x)\,dx = -x\cos(x) + \sin(x) + C$$

Example 4: Find the volume of the solid formed when the region bounded by $y = \cos(x)$, $x = 0$, and $x = \int x3^{-x}\,dx$ is rotated about the y-axis.

Solution:

1. It is easier to use the cylindrical shell method since we are rotating around a vertical line.

2. The radius of each shell will be x and the height of each shell will be $\cos(x)$.

3. The volume of the solid is given by $2\pi\int_0^{\pi/2} x\cos(x)\,dx$.

4. Use integration by parts to find the antiderivative for $x\cos(x)$. Let $u = x$ and $dv = \cos(x)dx$.

5. This results in $du = dx$ and $v = \sin(x)$.

6. Consequently, $\int x\cos(x)\,dx = x\sin(x) - \int \sin(x)\,dx = x\sin(x) + \cos(x)$.

$$2\pi\int_0^{\pi/2} x\cos(x)\,dx = 2\pi\left(x\sin(x) + \cos(x)\right)\Big|_0^{\pi/2} = 2\pi\left[\left(\left(\tfrac{\pi}{2}\right)\sin\left(\tfrac{\pi}{2}\right) + \cos\left(\tfrac{\pi}{2}\right)\right) - \left(0 + \cos(0)\right)\right]$$
$$= 2\pi\left(\tfrac{\pi}{2} - 1\right)$$

Example 5: Evaluate $\int xe^x\,dx$.

Solution: Let:

$u = x \qquad dv = e^x dx$

$du = dx \qquad v = e^x$

$\int xe^x\,dx = xe^x - \int e^x\,dx = xe^x - e^x + C$

Example 6: Evaluate $\int \ln(x)\,dx$.

Solution: There is only one function here—or is there? If we treat u as $\ln(x)$, then dv is dx and we get something interesting:

$u = \ln(x) \qquad dv = dx$

$du = \tfrac{1}{x}\,dx \qquad v = x$

$\int \ln(x)\,dx = x\ln(x) - \int \tfrac{1}{x}\,dx = x\ln(x) - x + C$

You now know the antiderivative of $\ln(x)$!

Example 7: Evaluate $\int x^3 e^{-2x} dx$.

Solution: Let's use the Tabular Method for this one. Let $u = x^3$ and $dv = e^{-2x}$.

u	dv	+1
x^3	e^{-2x}	1
$3x^2$	$\frac{-1}{2}e^{-2x}$	-1
$6x$	$\frac{1}{4}e^{-2x}$	1
6	$\frac{-1}{8}e^{-2x}$	-1
0	$\frac{1}{16}e^{-2x}$	1
0	$\frac{-1}{32}e^{-2x}$	-1

Figure 6.2
The Tabular Method for performing integration by parts on $\int x^3 e^{-2x} dx$.

$$\int x^3 e^{-2x} dx = \left(x^3\right)\left(\tfrac{-1}{2}e^{-2x}\right)(1) + \left(3x^2\right)\left(\tfrac{1}{4}e^{-2x}\right)(-1) + (6x)\left(\tfrac{-1}{8}e^{-2x}\right)(1) + (6)\left(\tfrac{1}{16}e^{-2x}\right)(-1) + C.$$

Therefore, $\int x^3 e^{-2x} dx = \tfrac{-1}{2}x^3 e^{-2x} - \tfrac{3}{4}x^2 e^{-2x} - \tfrac{3}{4}xe^{-2x} - \tfrac{3}{8}e^{-2x} + C$.

Example 8: Evaluate $\int x \ln(x) dx$.

Solution:

Let $u = x$ and $dv = \ln(x)$. This gives $du = dx$, which is fine, but $v = x \ln(x) - x$ is interesting. Let's see where this goes:

$$\int x \ln(x) dx = x(x \ln(x) - x) - \int x \ln(x) - x \, dx$$
$$\int x \ln(x) dx = x(x \ln(x) - x) - \int x \ln(x) dx + \int x \, dx$$

This is interesting—we have $\int x \ln(x) dx$ on both sides of the equation. Move the term from the right side of the equation to the left side and evaluate $\int x \, dx$:

$$2 \int x \ln(x) dx = x^2 \ln(x) - x^2 + \tfrac{1}{2}x^2$$

Divide by 2 to get the result $\int x \ln(x) dx = \tfrac{1}{2}x^2 \ln(x) - \tfrac{1}{4}x^2 + C$.

What would have happened if we did the last problem differently? Let $u = \ln(x)$ and $dv = x$. Then $du = \frac{1}{x}dx$ and $v = \frac{1}{2}x^2$. Make the substitution for integration by parts: $\int x \ln(x)\,dx = \frac{1}{2}x^2 \ln(x) - \int \frac{1}{2}x\,dx = <\frac{1}{2}x^2 \ln(x) - \frac{1}{4}x^2 + C$. It is not all that often that one can interchange functions when doing integration by parts and be able to reach an answer (and the same answer at that!).

YOU'VE GOT PROBLEMS

Problem 3: Evaluate $\int \sqrt{x} \ln(x)\,dx$.

Thus far, we've only concentrated on indefinite integrals in this section. The reason for doing so is that once the antiderivative is found, it is simply a matter of computing $F(b) - F(a)$ to get the definite integral. That is, $\int_1^2 x \ln(x)\,dx = \frac{1}{2}x^2 \ln(x) - \frac{1}{4}x^2 \Big|_1^2 = \left(2\ln(2) - 1\right) - \left(0 - \frac{1}{4}\right) = 2\ln(2) - \frac{3}{4}$.

Example 9: Let R be the region bounded by the functions $f(x) = \cos\left(\frac{\pi x}{2}\right)$, $g(x) = x^3 - 5x^2$, and the y-axis, as shown in the following diagram.

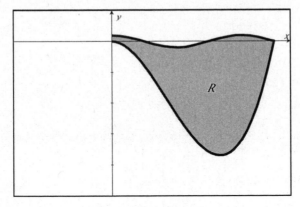

Figure 6.3
The region formed by the intersection of $f(x) = \cos\left(\frac{\pi x}{2}\right)$ *and* $g(x) = x^3 - 5x^2$.

1. Compute the area of R.

2. Determine the volume of the solid formed if R is rotated about the y-axis.

3. Determine the volume of the solid formed if R is rotated about the line $y = 1$.

Solution:

1. The area of the figure is $\int_0^5 \cos\left(\frac{\pi x}{2}\right) - \left(x^3 - 5x^2\right)dx = \frac{2}{\pi}\sin\left(\frac{\pi x}{2}\right) - \frac{1}{4}x^4 + \frac{5}{3}x^3\Big|_0^5 =$
   $\left(\frac{2}{\pi}\sin\left(\frac{5\pi}{2}\right) - \frac{1}{4}(5)^4 + \frac{5}{3}(5)^3\right) - 0 = \frac{2}{\pi} - \frac{625}{4} + \frac{625}{3} = \frac{625}{12} + \frac{2}{\pi}$ (or 52.720).

2. Using cylindrical shells with radius $= x$ and height $= \cos\left(\frac{\pi x}{2}\right) - \left(x^3 - 5x^2\right)$, the volume is
   $2\pi\int_0^5 x\left(\cos\left(\frac{\pi x}{2}\right) - x^3 + 5x^2\right)dx$.

3. The antiderivative for $x\cos\left(\frac{\pi x}{2}\right)$ will have to be done by parts. Let $u = x$ and $dv = \cos\left(\frac{\pi x}{2}\right)dx$, so that $du = dx$ and $v = \frac{2}{\pi}\sin\left(\frac{\pi x}{2}\right)$.

4. $\int x\cos\left(\frac{\pi x}{2}\right)dx = \frac{2}{\pi}x\sin\left(\frac{\pi x}{2}\right) - \frac{2}{\pi}\int\sin\left(\frac{\pi x}{2}\right)dx = \frac{2}{\pi}x\sin\left(\frac{\pi x}{2}\right) + \frac{4}{\pi^2}\cos\left(\frac{\pi x}{2}\right)$. Therefore,
   $2\pi\int_0^5 x\left(\cos\left(\frac{\pi x}{2}\right) - x^3 + 5x^2\right)dx = 2\pi\left(\frac{2}{\pi}x\sin\left(\frac{\pi x}{2}\right) + \frac{4}{\pi^2}\cos\left(\frac{\pi x}{2}\right) - \frac{1}{5}x^5 + \frac{5}{4}x^4\right)\Big|_0^5 =$
   $2\pi\left[\left(\frac{2}{\pi}(5)\sin\left(\frac{5\pi}{2}\right) + \frac{4}{\pi^2}\cos\left(\frac{5\pi}{2}\right) - \frac{1}{5}(5)^5 + \frac{5}{4}(5)^4\right) - \left(\frac{2}{\pi}(0)\sin(0) + \frac{4}{\pi^2}\cos(0)\right)\right] =$
   $2\pi\left[\frac{10}{\pi} + 0 - 625 + \frac{3125}{4} - \frac{4}{\pi^2}\right] = 2\pi\left[\frac{10}{\pi} + \frac{3125}{20} - \frac{4}{\pi^2}\right] = 20 + \frac{3125\pi}{10} - \frac{8}{\pi}$ (or 999.201).

5. A cross section of R will be a washer.

6. The radius of the larger circle will be $1 - (x^3 - 5x^2)$ while the radius of the smaller circle will be $1 - \cos\left(\frac{\pi x}{2}\right)dx$.

7. The volume of the solid is $\pi\int_0^5 1 - \left(x^3 - 5x^2\right)^2 - \left(1 - \cos\left(\frac{\pi x}{2}\right)\right)^2 dx$. As much fun as it would be to work out this integral, let's use the graphing calculator to get the answer.

$$\pi\cdot\int_0^5\left(\left(1 - x^3 + 5\cdot x^2\right)^2 - \left(1 - \cos\left(\frac{\pi\cdot x}{2}\right)\right)^2\right)dx \qquad \frac{4\cdot\left(1480\cdot\pi + 7\right)}{7}$$

$$1.\cdot\frac{4\cdot\left(1480\cdot\pi + 7\right)}{7} \qquad\qquad\qquad\qquad 2660.8898$$

Figure 6.4

A screen shot from a TI-Nspire computing the exact and approximate volumes of the solid formed when the region formed by the intersection of $f(x) = \cos\left(\frac{\pi x}{2}\right)dx$ *and* $g(x) = x^3 - 5x^2$ *is revolved about the line* $y = 1$.

Example 10: The function f(x) = 2sin(πx) is graphed over the interval [0,1]. Determine the volume of the solid formed when this region is rotated about:

1. The y-axis

2. The line $x = 1$

Solution:

1. Use cylindrical shells with radius = x and height = 2sin(πx).

2. The volume of the solid formed is $2\pi\int_0^1 2x\sin(\pi x)\,dx$.

3. Use integration by parts with $u = 2x$ and $dv = \sin(\pi x)\,dx$.

4. This makes $du = 2dx$ and $v = \frac{-1}{\pi}\cos(\pi x)$.

5. $2\pi\int_0^1 2x\sin(\pi x)\,dx = \frac{-2}{\pi}x\cos(\pi x) - \frac{-2}{\pi}\int\cos(\pi x)\,dx =$

   $\frac{-2}{\pi}x\cos(\pi x) + \frac{2}{\pi^2}\sin(\pi x)$. Therefore, $2\pi\int_0^1 2x\sin(\pi x)\,dx =$

   $2\pi\left[\left(\frac{-2}{\pi}(1)\cos(\pi) + \frac{2}{\pi^2}\sin(\pi)\right) - \left(\frac{-2}{\pi}(0)\cos(0) + \frac{2}{\pi^2}\sin(0)\right)\right] = 2\pi\left(\frac{2}{\pi}\right) = 4$.

6. Use cylindrical shells with radius = $1 - x$ and height = 2sin(πx).

7. The volume of the solid formed is $2\pi\int_0^1 (1-x)2\sin(\pi x)\,dx =$

   $2\pi\int_0^1 2\sin(\pi x) - 2x\sin(\pi x)\,dx = 2\pi\int_0^1 2\sin(\pi x)\,dx - 2\pi\int_0^1 2x\sin(\pi x)\,dx$.

   $2\pi\int_0^1 2\sin(\pi x)\,dx = -4\cos(\pi x)\Big|_0^1 = -4(\cos(\pi) - \cos(0)) = -4\,(-1-1) = 8$

8. Therefore, $2\pi\int_0^1 2\sin(\pi x)\,dx - 2\pi\int_0^1 2x\sin(\pi x)\,dx = 8 - 4 = 4$.

Example 11: The graph of g(x) = 3^{-x} is graphed over the interval [0,1]. Find the volume of the solid formed when this region is rotated about:

1. The y-axis

2. The line $x = -1$

Solution:

1. Use cylindrical shells with radius = x and height = 3^{-x}. The volume of the solid formed is

   $2\pi\int_0^1 x3^{-x}\,dx$.

2. Use integration by parts with $u = x$ and $dv = 3^{-x}$. This gives $du = dx$ and $v = \frac{-1}{\ln(3)}3^{-x}$.

   $2\pi\int_0^1 x3^{-x}\,dx = \frac{-x}{\ln(3)}3^{-x} - \frac{-1}{\ln(3)}\int 3^{-x}\,dx = \frac{-x}{\ln(3)}3^{-x} - \frac{1}{(\ln(3))^2}3^{-x}$

3. Therefore, $2\pi\int_0^1 x3^{-x}\,dx = 2\pi\left(\frac{-x}{\ln(3)}3^{-x} - \frac{1}{(\ln(3))^2}3^{-x}\right)\Big|_0^1 =$

$2\pi\left[\left(\frac{-1}{\ln(3)}3^{-1} - \frac{1}{(\ln(3))^2}3^{-1}\right) - \left(0 - \frac{1}{(\ln(3))^2}\right)\right] = 2\pi\left(\frac{-1}{3\ln(3)} - \frac{1}{3(\ln(3))^2} + \frac{1}{(\ln(3))^2}\right) = 2\pi\left(\frac{2-\ln(3)}{3(\ln(3))^2}\right).$

4. Use cylindrical shells with radius $r = x+1$ and height $= 3^{-x}$. The volume of the solid formed is $2\pi\int_0^1 (x+1)3^{-x}\,dx = 2\pi\int_0^1 x3^{-x}\,dx + 2\pi\int_0^1 3^{-x}\,dx = 2\pi\left(\frac{2-\ln(3)}{3(\ln(3))^2}\right) +$

$2\pi\left(\frac{-1}{\ln(3)}3^{-x}\right)\Big|_0^1 = 2\pi\left(\frac{2-\ln(3)}{3(\ln(3))^2}\right) + 2\pi\left(\frac{-1}{3\ln(3)} - \frac{-1}{\ln(3)}\right) = 2\pi\left(\frac{2-\ln(3)}{3(\ln(3))^2}\right) + 2\pi\left(\frac{2}{3\ln(3)}\right) =$

$2\pi\left(\frac{2-\ln(3)}{3(\ln(3))^2} + \frac{2\ln(3)}{3(\ln(3))^2}\right) = 2\pi\left(\frac{2+\ln(3)}{3(\ln(3))^2}\right).$

Two Transcendentals

The case of using integration by parts leads to some useful formulas for powers of the secant and tangent functions (and by extension, powers of cosecant and cotangent). We'll work our way into this.

Example 12: Evaluate $\int e^x \sin(x)dx$.

Solution:

1. Let $u = \sin(x)$ and $dv = e^x dx$. (I chose this order arbitrarily.) This gives $du = \cos(x)\,dx$ and $v = e^x$:

$$\int e^x \sin(x)dx = e^x \sin(x) - \int e^x \cos(x)dx$$

2. Let $u = \cos(x)$ and $dv = e^x dx$. This gives $du = -\sin(x)dx$ and $v = e^x$:

$$\int e^x \sin(x)dx = e^x \sin(x) - \int e^x \cos(x)dx = e^x \sin(x) - \left(e^x \cos(x) - \int -e^x \sin(x)\,dx\right) =$$
$$e^x\sin(x) - e^x \cos(x) - \int e^x \sin(x)dx$$

3. Add $\int e^x \sin(x)dx$ to both sides of the equation:

$$2<\int e^x \sin(x)dx = e^x \sin(x) - e^x \cos(x)$$

4. This gives the final answer to the problem:

$$\int e^x \sin(x)dx = \frac{e^x \sin(x) + e^x \cos(x)}{2} + C$$

Example 13: Evaluate $\int \sec^3(x)\,dx$.

Solution:

1. Rewrite the integrand as $\sec(x)$ and $\sec^2(x)$.

2. Let $u = \sec(x)$ and $du = \sec^2(x)\,dx$. This gives $du = \sec(x)\tan(x)\,dx$ and $v = \tan(x)$:

$$\int \sec^3(x)\,dx = \sec(x)\tan(x) - \int \sec(x)\tan^2(x)\,dx$$

3. Since $1 + \tan^2(x) = \sec^2(x)$, replace $\tan^2(x)$ with $\sec^2(x) - 1$:

$$\int \sec^3(x)\,dx = \sec(x)\tan(x) - \int \sec(x)\left(\sec^2(x) - 1\right)dx = \sec(x)\tan(x)$$
$$- \int \sec^3(x)\,dx + \int \sec(x)\,dx$$

4. Add $\int \sec^3(x)\,dx$ to both sides of the equation and recall that $\int \sec(x)\,dx = \ln|\sec(x) + \tan(x)|$:

$$2\int \sec^3(x)\,dx = \sec(x)\tan(x) + \ln|\sec(x) + \tan(x)|$$

5. Divide by 2:

$$\int \sec^3(x)\,dx = \tfrac{1}{2}\sec(x)\tan(x) + \tfrac{1}{2}\ln|\sec(x) + \tan(x)| + C$$

Example 14: Evaluate $\int \sec^5(x)\,dx$.

Solution:

1. Let $u = \sec^3(x)$ and $dv = \sec^2(x)dx$:

$du = 3\sec^2(x)\sec(x)\tan(x)\,dx = 3\sec^3(x)\tan(x)\,dx$ and $v = \tan(x)$

$$\int \sec^5(x)\,dx = \sec^3(x)\tan(x) - \int 3\sec^3(x)\tan^2(x)\,dx = \sec^3(x)\tan(x)$$
$$- \int 3\sec^3(x)\left(\sec^2(x) - 1\right)dx$$

2. Simplify the last integral:

$$\int \sec^5(x)\,dx = \sec^3(x)\tan(x) - 3\int \sec^5(x)\,dx + 3\int \sec^3(x)\,dx$$

3. Add $3\int \sec^5(x)\,dx$ to both sides of the equation:

$$4\int \sec^5(x)\,dx = \sec^3(x)\tan(x) + 3\int \sec^3(x)\,dx$$

4. We just evaluated $\int \sec^3(x)\,dx$ in Example 13 so we have:

$$4\int \sec^5(x)\,dx = \sec^3(x)\tan(x) + 3\left(\tfrac{1}{2}\sec(x)\tan(x) + \tfrac{1}{2}\ln|\sec(x) + \tan(x)|\right)$$

5. Divide by 4:

$$\int \sec^5(x)\,dx > = \tfrac{1}{4}\sec^3(x)\tan(x) + \tfrac{3}{8}\sec(x)\tan(x) + \tfrac{3}{8}\ln|\sec(x) + \tan(x)| + C$$

6. We could continue with other odd powers for sec(x). The result is:

$$\int \sec^n(x)\,dx = \tfrac{1}{n-1}\sec^{n-2}(x)\tan(x) + \tfrac{n-2}{n-1}\int \sec^{n-2}(x)\,dx$$

7. In other words, the process continues to create another integral with a reduced exponent. In the case of even powers, the last integral $\int \sec^2(x)\,dx = \tan(x)$.

CRITICAL POINT

If n is a positive odd integer, then

$$\int \sec^n(x)\,dx = \tfrac{1}{n-1}\sec^{n-2}(x)\tan(x) + \tfrac{n-2}{n-1}\int \sec^{n-2}(x)\,dx \,.$$

What happens if n is a positive even integer?

Example 15: Evaluate $\int \sec^6(x)\,dx$.

Solution:

1. Rewrite the integrand as $\sec^2(x)\sec^4(x)$. Rewrite $\sec^4(x)$ as $(\tan^2(x) + 1)^2$:

$$\int \sec^6(x)\,dx = \int \sec^2(x)\left(\tan^2(x)+1\right)^2 dx$$

2. Expand the quadratic $(\tan^2(x) + 1)^2$:

$$\int \sec^6(x)\,dx = \int \sec^2(x)\left(\tan^4(x)+2\tan^2(x)+1\right)dx$$

3. Distribute the $\sec^2(x)$ through this expansion:

$$\int \tan^4(x)\sec^2(x)+2\tan^2(x)\sec^2(x)+\sec^2(x)\,dx = \tfrac{1}{5}\tan^5(x)+\tfrac{2}{3}\tan^3(x)+\tan(x)+C$$

YOU'VE GOT PROBLEMS

Problem 4: Evaluate $\int e^{2x}\cos(3x)\,dx$.

The Least You Need to Know

- Integration by parts is a process for integrating functions that are the result of the product rule from differentiation.
- Integrate functions are the product of a polynomial function and a transcendental function.
- Integrate functions are the product of a two transcendental functions.
- The integration of $\int \sec''(x)\,dx$ is done with integrations by parts.

Integration with Trigonometric Functions

We continue to study patterns in integrands. In this chapter, the concentration will be on integrands with trigonometric functions and integrands that can be transformed to contain trigonometric functions. This might be a good time for you to refer to Chapter 1 and look over the trigonometric identities.

We'll begin with problems that lend themselves to right triangle trigonometry and then we'll look at products of trigonometric functions that are related to one another by the derivative.

In This Chapter

- Using trigonometric substitutions to deal with integrands of the forms $\sqrt{a^2 + x^2}, \sqrt{a^2 - x^2}, \sqrt{x^2 - a^2}$

- Investigating integrands of the form $\sin^n(x) \cos^m(x)$

- Investigating integrands of the form $\tan^n(x) \sec^m(x)$ (m is even)

- Investigating integrands of the form $\tan^n(x) \sec^m(x)$ (m is odd, n is even)

Trigonometric Substitutions

Most students who were successful at geometry can answer the question, "What is the Pythagorean Theorem?" Almost without fail, the answer will be, "$a^2 + b^2 = c^2$." This is a (practically) correct answer provided that both the person asking the question and the person answering the question understand that if the length of the hypotenuse is labeled as c, then this equation fits the Pythagorean Theorem. However, you need to be sure you identify which variable represents the hypotenuse of the right triangle. (Can you tell that I taught geometry for a number of years and would get a tad frustrated when a student tried to apply the formula to a triangle that wasn't a right triangle? But I digress.) We can take advantage of this special relationship and of right triangle trigonometry whenever the integrand of our problem is in the form $\sqrt{a^2 + x^2}$, $\sqrt{x^2 - a^2}$, or $\sqrt{a^2 - x^2}$.

Case I: $\sqrt{a^2 + x^2}$

If $a^2 + b^2 = c^2$, then $c = \sqrt{a^2 + b^2}$. We'll draw a right triangle with legs designated by a and x and mark the angle between the leg designated by a and the hypotenuse as θ.

Figure 7.1

Right triangle with legs of length a and x.

Using right trigonometry, $\tan(\theta) = \frac{x}{a}$ so that $x = a\tan(\theta)$ and $\sec(\theta) = \frac{\sqrt{a^2 + x^2}}{a}$ so that $\sqrt{a^2 + x^2} = a\sec(\theta)$. With $x = a\tan(\theta)$, $dx = a\sec^2(\theta)d\theta$.

Example 1: Evaluate $\int \sqrt{25 + x^2}\, dx$.

Solution: Let $x = 5\tan(\theta)$, $dx = 5\sec^2(\theta)d\theta$, and $\sqrt{25 + x^2} = 5\sec(\theta)$. When the integral is transformed, the problem is now $\int \left(5\sec(\theta)\right)\left(5\sec^2(\theta)d\theta\right) = 25\int \sec^3(\theta)\, d\theta$. As we saw at the end of Chapter 6, $\int \sec^3(\theta)\, dx = \frac{1}{2}\sec(\theta)\tan(\theta) + \frac{1}{2}\ln|\sec(\theta) + \tan(\theta)| + C$.

Therefore, $\int \sqrt{25 + x^2}\, dx = 25\left(\frac{1}{2}\left(\frac{x}{5}\right)\left(\frac{\sqrt{25 + x^2}}{5}\right) + \frac{1}{2}\ln\left|\frac{\sqrt{25 + x^2}}{5} + \frac{x}{5}\right|\right) + C =$

$25\left[\frac{x\sqrt{25 + x^2}}{50} + \frac{1}{2}\ln\left|\frac{\sqrt{25 + x^2} + x}{5}\right|\right] + C = \frac{x\sqrt{25 + x^2}}{2} + \frac{25}{2}\ln\left|\frac{\sqrt{25 + x^2} + x}{5}\right| + C$.

If you worked that problem through on paper and looked at this result, I'll bet yours is different. Here's the reason why:

1. The term $\ln\left|\frac{\sqrt{25+x^2}}{5} + \frac{x}{5}\right| = \ln\left|\frac{\sqrt{25+x^2}+x}{5}\right| = \ln\left|\sqrt{25+x^2} + x\right| - \ln(5)$.

2. When you multiply the constant $\ln(5)$ by the constant $\frac{25}{2}$, the answer is still a constant.

3. When this constant is added to the constant of integration, C, the answer is still a constant.

4. So the extra term that you had is absorbed into the constant of integration.

BE AWARE

Don't confuse the result for $\int \sqrt{25+x^2}\, dx$ for a problem such as $\int x\sqrt{25+x^2}\, dx$. The u-substitution method applies to this problem because the derivative of $25 + x^2$ is $2x$ and the introduction of the factors 2 and $\frac{1}{2}$ results in the problem becoming $\frac{1}{2}\int \sqrt{u}\, du$.

Example 2: Evaluate $\int \sqrt{5+4z^2}\, dz$.

Solution: Let $2z = \sqrt{5}\tan(\theta)$, $dz = \frac{\sqrt{5}}{2}\sec^2(\theta)\, d\theta$, and $\sqrt{5+4z^2} = \sqrt{5}\sec(\theta)$. When the integral is transformed, the problem becomes $\int \left(\sqrt{5}\sec(\theta)\right)\left(\frac{\sqrt{5}}{2}\sec^2(\theta)d\theta\right) = \frac{5}{2}\int \sec^3(\theta)\, d\theta$:

$$\int \sqrt{5+4z^2}\, dz = \frac{5}{2}\int \sec^3(\theta)\, d\theta = \frac{5}{2}\left(\frac{1}{2}\sec(\theta)\tan(\theta) + \frac{1}{2}\ln\left|\sec(\theta) + \tan(\theta)\right|\right) + C =$$

$$\frac{5}{4}\left(\frac{\sqrt{5+4z^2}}{\sqrt{5}}\right)\left(\frac{2z}{\sqrt{5}}\right) + \frac{5}{4}\ln\left|\frac{\sqrt{5+4z^2}}{\sqrt{5}} + \frac{2z}{\sqrt{5}}\right| + C$$

$$\int \sqrt{5+4z^2}\, dz = \frac{z\sqrt{5+4z^2}}{2} + \frac{5}{4}\ln\left|\sqrt{5+4z^2} + 2z\right| + C$$

YOU'VE GOT PROBLEMS

Problem 1: Evaluate $\int_0^{10} \sqrt{49+25x^2}\, dx$.

Case II: $\sqrt{x^2 - a^2}$

The hypotenuse of the right triangle is designated as *x*, and one leg is labeled *a*. Let the acute angle θ be between the leg *a* and the hypotenuse. The only reason for putting θ in this location is that we get to avoid dragging negative signs through the problem. Otherwise, there is no difference in the process.

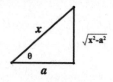

Figure 7.2
Right triangle with hypotenuse with length x and leg with length a.

1. Looking at the diagram for this problem, we get $\cos(\theta) = \frac{a}{x}$ so $\sec(\theta) = \frac{x}{a}$ and $x = a\sec(\theta)$ while $\tan(\theta) = \frac{\sqrt{x^2 - a^2}}{a}$, which gives $\sqrt{x^2 - a^2} = a\tan(\theta)$.

2. The differential $dx = a\sec(\theta)\tan(\theta)d\theta$.

3. Transform the integral $\int \sqrt{x^2 - a^2}\, dx$ to $\int \left(a\tan(\theta) \right)\left(a\sec(\theta)\tan(\theta)d\theta \right) = a^2 \int \sec(\theta)\tan^2(\theta)\, d\theta$.

4. Use the trigonometric identity $\tan^2(\theta) = \sec^2(\theta) - 1$.

5. The integral becomes $a^2 \int \sec(\theta)\left(\sec^2(\theta) - 1 \right)d\theta = a^2 \int \sec^3(\theta) - \sec(\theta)d\theta$.
   Since $\int \sec^3(\theta)d\theta = \frac{1}{2}\sec(\theta)\tan(\theta) + \frac{1}{2}\ln|\sec(\theta) + \tan(\theta)| + C$ and
   $\int \sec(\theta)d\theta = \ln|\sec(\theta) + \tan(\theta)| + C$:

   $$a^2 \int \sec^3(\theta) - \sec(\theta)d\theta = a^2 \left(\frac{1}{2}\sec(\theta)\tan(\theta) + \frac{1}{2}\ln|\sec(\theta) + \tan(\theta)| - \ln|\sec(\theta) + \tan(\theta)| \right) + C$$
   $$= a^2 \left(\frac{1}{2}\sec(\theta)\tan(\theta) - \frac{1}{2}\ln|\sec(\theta) + \tan(\theta)| \right) + C$$

6. Transform the integral back to the original variables:

   $$\int \sqrt{x^2 - a^2}\, dx = a^2 \left(\frac{1}{2}\left(\frac{x}{a}\right)\left(\frac{\sqrt{x^2 - a^2}}{a}\right) - \frac{1}{2}\ln\left|\frac{x}{a} + \frac{\sqrt{x^2 - a^2}}{a}\right| \right) + C = \frac{x\sqrt{x^2 - a^2}}{2} - \frac{a^2 \ln\left|x + \sqrt{x^2 - a^2}\right|}{2} + C$$

Example 3: Evaluate $\int \sqrt{x^2 - 25}\, dx$.

Solution: $\int \sqrt{x^2 - 25}\, dx = \dfrac{x\sqrt{x^2-25}}{2} - \dfrac{25\ln\left|x + \sqrt{x^2-25}\right|}{2} + C$.

BE AWARE

If you feel like you need to memorize these results, I would suggest that you memorize $\int \sec^3(\theta)\, d\theta$ and $\int \sec(\theta)\, d\theta$. In this way, you do not have to concern yourself between the slight variations of the $\sqrt{a^2 + x^2}$ and $\sqrt{x^2 - a^2}$ formulas.

Example 4: Evaluate $\int \sqrt{25x^2 - 9}\, dx$.

Solution:

1. Factor 25 from the radicand to get $\sqrt{25x^2 - 9} = \sqrt{25\left(x^2 - \frac{9}{25}\right)} = 5\sqrt{\left(x^2 - \frac{9}{25}\right)}$:

    $$5\int \sqrt{\left(x^2 - \tfrac{9}{25}\right)}\, dx = 5\left(\dfrac{x\sqrt{x^2 - \frac{9}{25}}}{2} - \dfrac{\frac{9}{25}\ln\left|x + \sqrt{x^2 - \frac{9}{25}}\right|}{2} \right) + C = \dfrac{x\sqrt{25x^2 - 9}}{2} - \dfrac{9\ln\left|x + \sqrt{x^2 - \frac{9}{25}}\right|}{10} + C$$

2. Let's do a little work with the expression $x + \sqrt{x^2 - \frac{9}{25}}$. First, get a common denominator within the radical, $x + \sqrt{\frac{25x^2 - 9}{25}}$.

3. Simplify the radical, $x + \dfrac{\sqrt{25x^2 - 9}}{5}$.

4. Get a common denominator for the entire term, $\dfrac{5x + \sqrt{25x^2 - 9}}{5}$.

5. Back to the antiderivative, $\dfrac{x\sqrt{25x^2 - 9}}{2} - \dfrac{9\ln\left|x + \sqrt{x^2 - \frac{9}{25}}\right|}{10} + C = \dfrac{x\sqrt{25x^2 - 9}}{2} - \dfrac{9\ln\left|5x + \sqrt{25x^2 - 9}\right|}{10} + C$
    (with the $-\ln(5)$ included in the constant of integration).

YOU'VE GOT PROBLEMS

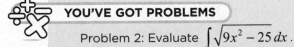

Problem 2: Evaluate $\int \sqrt{9x^2 - 25}\, dx$.

Case III: $\sqrt{a^2 - x^2}$

The length of the hypotenuse is a and the length of one of the legs is x. The acute angle opposite x is θ. The only reason for putting θ in this location is that we get to avoid dragging negative signs through the problem. Otherwise, there is no difference in the process.

Figure 7.3

Right triangle with hypotenuse with length a and leg with length x.

Looking at the diagram for this problem, we get $\sin(\theta) = \frac{x}{a}$ so $x = a\sin(\theta)$ and $dx = a\cos(\theta)$ $d\theta$. $\cos(\theta) = \frac{\sqrt{a^2 - x^2}}{a}$ so that $\sqrt{a^2 - x^2} = a\cos(\theta)$. Transform the integral $\int \sqrt{a^2 - x^2}\, dx$ to $\int (a\cos(\theta))(a\cos(\theta)d\theta) = a^2 \int \cos^2(\theta)d\theta$. Manipulate the trigonometric identity $\cos(2x) = 2\cos2(x) - 1$ to $\cos^2(x) = \frac{\cos(2x)+1}{2}$.

$a^2 \int \cos^2(\theta)d\theta = \frac{a^2}{2}\int \cos(2\theta) + 1 d\theta = \frac{a^2}{2}\left(\frac{1}{2}\sin(2\theta) + \theta\right) + C = \frac{a^2}{2}\left(\sin(\theta)\cos(\theta) + \theta\right) + C$.

(Recall that $\sin(2x) = 2\sin(x)\cos(x)$.)

Transform the result back to the original variable:

$$\frac{a^2}{2}\left(\sin(\theta)\cos(\theta) + \theta\right) + C = \frac{a^2}{2}\left(\left(\frac{x}{a}\right)\left(\frac{\sqrt{a^2 - x^2}}{a}\right) + \sin^{-1}\left(\frac{x}{a}\right)\right) + C =$$
$$\frac{1}{2}x\sqrt{a^2 - x^2} + \frac{a^2}{2}\sin^{-1}\left(\frac{x}{a}\right) + C$$

Example 5: Evaluate $\int \sqrt{100 - x^2}\, dx$.

Solution: $\int \sqrt{100 - x^2}\, dx = \frac{1}{2}x\sqrt{100 - x^2} + 50\sin^{-1}\left(\frac{x}{10}\right) + C$.

Example 6: Evaluate $\int \sqrt{100 - 49x^2}\, dx$:

$$\int \sqrt{100 - 49x^2}\, dx = 7\int \sqrt{\frac{100}{49} - x^2}\, dx = 7\left(\frac{1}{2}x\sqrt{\frac{100}{49} - x^2} + \frac{50}{49}\sin^{-1}\left(\frac{7x}{10}\right)\right) + C =$$
$$\frac{1}{2}x\sqrt{100 - 49x^2} + \frac{50}{7}\sin^{-1}\left(\frac{7x}{10}\right) + C$$

YOU'VE GOT PROBLEMS

Problem 3: Evaluate $\int \sqrt{49 - 100x^2} \, dx$.

Integrals of the Form $\sin^n(x) \cos^m(x)$ (When Either m or n Is Odd)

Integrands of the form $\sin^n(x) \cos(x)$ or $\cos^n(x) \sin(x)$ are easily solved using u-substitution. What happens if both the sine function and cosine function are raised to a power? There are three cases to consider—both exponents are even, both are odd, or one of them is even and the other is odd.

Case I: Both m and n Are Odd

This turns out to be a pretty simple case. We'll factor a $\cos(x)$ from the cosine term and rewrite the remaining term as a binomial in the form $\left(1 - \sin^2(x)\right)^{n/2}$ and then use binomial expansion.

Example 7: Evaluate $\int \sin^3(x) \cos^5(x) \, dx$.

Solution: Rewrite $\int \sin^3(x) \cos^5(x) \, dx$ as $\int \sin^3(x) \cos^4(x) \cos(x) \, dx =$

$\int \sin^3(x) \left(\cos^2(x)\right)^2 \cos(x) \, dx = \int \sin^3(x) \left(1 - \sin^2(x)\right)^2 \cos(x) \, dx$.

Expand the binomial: $(1 - \sin^2(x))^2 = 1 - 2\sin^2(x) + \sin^4(x)$. Substitute this into the integrand and distribute:

$$\int \sin^3(x)\left(1 - 2\sin^2(x) + \sin^4(x)\right) \cos(x) \, dx =$$

$$\int \sin^3(x)\cos(x) - 2\sin^5(x)\cos(x) + \sin^7(x)\cos(x) \, dx =$$

$$\tfrac{1}{4}\sin^4(x) - \tfrac{1}{3}\sin^6(x) + \tfrac{1}{8}\sin^8(x) + C$$

CRITICAL POINT

The expansion of $(a + b)^n = \displaystyle\sum_{r=0}^{n} \begin{pmatrix} n \\ r \end{pmatrix} a^{n-r} b^r$. $\begin{pmatrix} n \\ r \end{pmatrix}$ is the number of combinations of n items taken r at a time. It is computed by multiplying the first r integers, counting down from n, and dividing this by the first r integers, counting from 1. For example, $\begin{pmatrix} 7 \\ 3 \end{pmatrix} = \frac{7 \times 6 \times 5}{1 \times 2 \times 3} = 35$ while $\begin{pmatrix} 6 \\ 4 \end{pmatrix} = \frac{6 \times 5 \times 4 \times 3}{1 \times 2 \times 3 \times 4} = 15$. By definition, $\begin{pmatrix} n \\ 0 \end{pmatrix} = 1$.

Example 8: Evaluate $\int \sin^7(x)\cos^{11}(x)\,dx$.

Solution:

$$\int \sin^7(x)\cos^{11}(x)\,dx = \int \sin^6(x)\sin(x)\cos^{11}(x)\,dx = \int \left(\sin^2(x)\right)^3 \sin(x)\cos^{11}(x)\,dx =$$
$$\int \left(1-\cos^2(x)\right)^3 \sin(x)\cos^{11}(x)\,dx$$

$$\left(1-\cos^2(x)\right)^3 = 1-3\cos^2(x)+3\cos^4(x)-\cos^6(x)$$

$$\int \left(1-\cos^2(x)\right)^3 \sin(x)\cos^{11}(x)\,dx =$$

$$\int \left(1-3\cos^2(x)+3\cos^4(x)-\cos^6(x)\right)\sin(x)\cos^{11}(x)\,dx$$

$$\int \left(1-\cos^2(x)\right)^3 \sin(x)\cos^{11}(x)\,dx =$$
$$\int \sin(x)\cos^{11}(x)-3\sin(x)\cos^{13}(x)+3\sin(x)\cos^{15}(x)-\sin(x)\cos^{17}(x)\,dx$$

$$\int \left(1-\cos^2(x)\right)^3 \sin(x)\cos^{11}(x)\,dx =$$
$$\tfrac{-1}{12}\cos^{12}(x)+\tfrac{3}{14}\cos^{14}(x)-\tfrac{3}{16}\cos^{16}(x)-\tfrac{1}{18}\cos^{18}(x)+C$$

YOU'VE GOT PROBLEMS

Problem 4: Evaluate $\int \sin^9(x)\cos^9(x)\,dx$.

Case II: Both m and n Are Even

This is going to get sloppy—there are no two ways about it. We'll use $\sin^2(x)=\frac{1-\cos(2x)}{2}$ and $\cos^2(x)=\frac{1+\cos(2x)}{2}$.

Example 9: Evaluate $\int \sin^2(x)\cos^2(x)\,dx$.

Solution: Replace each of the trigonometric functions with the preceding identities to get

$$\int \sin^2(x)\cos^2(x)\,dx = \int \left(\tfrac{1-\cos(2x)}{2}\right)\left(\tfrac{1+\cos(2x)}{2}\right)dx = \tfrac{1}{4}\int 1-\cos^2(2x)\,dx.$$

Use the identity again:

$$\tfrac{1}{4}\int 1-\cos^2(2x)\,dx \;=\; \tfrac{1}{4}\int 1-\tfrac{1+\cos(4x)}{2}\,dx \;=\; \tfrac{1}{4}\int \tfrac{2-(1+\cos(4x))}{2}\,dx \;=\; \tfrac{1}{8}\int 1-\cos(4x)\,dx \;=$$

$$\tfrac{1}{8}x-\tfrac{1}{32}\sin(4x)+C$$

Now we have the chance to really get involved (algebraically speaking, of course).

Example 10: Evaluate $\int \sin^4(x)\cos^6(x)\,dx$.

Solution: Use the identities for $\sin^2(x)$ and $\cos^2(x)$ to get:

$$\int \sin^4(x)\cos^6(x)\,dx \;=\; \int \left(\sin^2(x)\right)^2 \left(\cos^2(x)\right)^3 dx \;=\; \int \left(\tfrac{1-\cos(2x)}{2}\right)^2 \left(\tfrac{1+\cos(2x)}{2}\right)^3 dx$$

At this point, we are not going to subject ourselves to unnecessary pain by trying to slog through all that algebra. Now that you understand the principle behind how the problem is done, let's use that old-fashioned phrase "Details are left to the reader" and use our calculator to learn that:

$$\int \sin^4(x)\cos^6(x)\,dx \;=$$

$$\tfrac{-1}{10}\sin^3(x)\cos^7(x) - \tfrac{3}{80}\sin(x)\cos^7(x) + \tfrac{1}{160}\sin(x)\cos^5(x) + \tfrac{1}{128}\sin(x)\cos^3(x) +$$

$$\tfrac{3}{256}\sin(x)\cos(x) + \tfrac{3}{256}x + C$$

And kudos to any of you who do all that work!

Case III: Either m or n Is Odd

This case is handled in the same way as in Case I.

Example 11: Evaluate $\int \sin^3(x)\cos^4(x)\,dx$.

Solution: Rewrite $\sin^3(x)$ as $\sin^2(x)\sin(x)$ and then use the Pythagorean identity:

$$\int \sin^3(x)\cos^4(x)\,dx \;=\; \int \sin^2(x)\sin(x)\cos^4(x)\,dx \;=\; \int \left(1-\cos^2(x)\right)\sin(x)\cos^4(x)\,dx \;=$$

$$\int \cos^4(x)\sin(x) - \cos^6(x)\sin(x)\,dx$$

$$\int \sin^3(x)\cos^4(x)\,dx \;=\; \tfrac{1}{7}\cos^7(x) - \tfrac{1}{5}\cos(x) + C$$

Integrals with Integrands of the Form $\tan^n(x)$ $\sec^m(x)$ (*m* Is Even)

This case is also pretty straightforward. Because sec(x) is being raised to an even power, we'll rewrite $\sec^n(x)$ as $\sec^{n-2}(x)\sec^2(x)$. We will then apply the Pythagorean identity to $\sec^{n-2}(x)$ and rewrite it as $(1 + \tan^2(x))^{n-2}$, apply the binomial expansion, distribute $\tan^n(x)\sec^2(x)$, and integrate.

Example 12: Evaluate $\int \tan^5(x)\sec^6(x)\,dx$.

Solution:

1. Rewrite $\int \tan^5(x)\sec^6(x)\,dx$ as $\int \tan^5(x)\sec^4(x)\sec^2(x)\,dx$.

2. Apply the identity: $\int \tan^5(x)\left(1+\tan^2(x)\right)^2\sec^2(x)\,dx$.

3. Expand the binomial: $\int \tan^5(x)\left(1+2\tan^2(x)+\tan^4(x)\right)\sec^2(x)\,dx$.

4. Distribute: $\int \tan^5(x)\sec^2(x)+2\tan^7(x)\sec^2(x)+\tan^9(x)\sec^2(x)\,dx$.

5. Integrate: $\int \tan^5(x)\sec^6(x)\,dx = \frac{1}{6}\tan^6(x)+\frac{1}{4}\tan^8(x)+\frac{1}{10}\tan^{10}(x)+C$.

Example 13: Evaluate $\int \tan^4(3x)\sec^8(3x)\,dx$.

Solution:

1. Rewrite: $\int \tan^4(3x)\sec^8(3x)\,dx = \int \tan^4(3x)\sec^6(3x)\sec^2(3x)\,dx =$
$\int \tan^4(3x)\left(\sec^2(3x)\right)^3\sec^2(3x)\,dx$.

2. Apply the identity: $\int \tan^4(3x)\left(1+\tan^2(3x)\right)^3\sec^2(3x)\,dx$.

3. Expand the binomial: $\int \tan^4(3x)\left(1+3\tan^2(3x)+3\tan^4(3x)+\tan^6(3x)\right)\sec^2(3x)\,dx$.

4. Distribute:
$\int \tan^4(3x)\sec^2(3x)+3\tan^6(3x)\sec^2(3x)+3\tan^8(3x)\sec^2(3x)+\tan^{10}(3x)\sec^2(3x)\,dx$.

5. Integrate: $\int \tan^4(3x)\sec^8(3x)\,dx = \frac{1}{15}\tan^5(3x)+\frac{1}{7}\tan^7(3x)+\frac{1}{9}\tan^9(x)+\frac{1}{33}\tan^{11}(x)+C$.

YOU'VE GOT PROBLEMS

Problem 5: Evaluate $\int \tan^2(5x)\sec^{10}(5x)\,dx$.

Integrals with Integrands of the Form $\tan^n(x)$ $\sec^m(x)$ (m Is Odd, n Is Even)

These problems are fairly straightforward, but they get intense in their use of algebra. We'll take advantage, once again, of the Pythagorean identity $\tan^2(x) = \sec^2(x) - 1$ to get an integrand that is entirely in $\sec(x)$. The problem with this is that it requires we use the reduction formula for $\int \sec^n(x)\,dx$.

Example 14: Evaluate $\int \tan^4(x)\sec^3(x)\,dx$.

Solution:

1. Apply the identity: $\int \tan^4(x)\sec^3(x)\,dx = \int \left(\tan^2(x)\right)^2 \sec^3(x)\,dx = \int \left(\sec^2(x) - 1\right)^2 \sec^3(x)\,dx$

2. Expand the binomial: $\int \left(\sec^4(x) - 2\sec^2(x) + 1\right)\sec^3(x)\,dx$.

3. Distribute: $\int \sec^7(x) - 2\sec^5(x) + \sec^3(x)\,dx$.

4. Use the result from Chapter 6 for $\int \sec^3(x)\,dx$:

   $\int \sec^3(x)\,dx = \frac{1}{2}\sec(x)\tan(x) + \frac{1}{2}\ln\left|\sec(x) + \tan(x)\right|$

5. Use the reduction formula for the higher powers of $\sec(x)$:

   $\int \sec^n(x)\,dx = \frac{1}{n-1}\sec^{n-2}(x)\tan(x) + \frac{n-2}{n-1}\int \sec^{n-2}(x)\,dx$:

   $\int \sec^5(x)\,dx = \frac{1}{4}\sec^3(x)\tan(x) + \frac{3}{4}\int \sec^3(x)\,dx$

   $= \frac{1}{4}\sec^3(x)\tan(x) + \frac{3}{4}\left(\frac{1}{2}\sec(x)\tan(x) + \frac{1}{2}\ln\left|\sec(x) + \tan(x)\right|\right)$

   $= \frac{1}{4}\sec^3(x)\tan(x) + \frac{3}{8}\sec(x)\tan(x) + \frac{3}{8}\ln\left|\sec(x) + \tan(x)\right|$

   $\int \sec^7(x)\,dx = \frac{1}{6}\sec^5(x)\tan(x) + \frac{5}{6}\int \sec^5(x)\,dx$

   $= \frac{1}{6}\sec^5(x)\tan(x) + \frac{5}{6}\left(\frac{1}{4}\sec^3(x)\tan(x) + \frac{3}{8}\sec(x)\tan(x) + \frac{3}{8}\ln\left|\sec(x) + \tan(x)\right|\right)$

   $= \frac{1}{6}\sec^5(x)\tan(x) + \frac{5}{24}\sec^3(x)\tan(x) + \frac{5}{16}\sec(x)\tan(x) + \frac{5}{16}\ln\left|\sec(x) + \tan(x)\right|$

6. Therefore, $2\left(\frac{1}{4}\sec^3(x)\tan(x) + \frac{3}{8}\sec(x)\tan(x) + \frac{3}{8}\ln\left|\sec(x)+\tan(x)\right|\right) =$

$$\int\tan^4(x)\sec^3(x)\,dx = \int\sec^7(x) - 2\sec^5(x) + \sec^3(x)\,dx =$$

$$\frac{1}{6}\sec^5(x)\tan(x) + \frac{5}{24}\sec^3(x)\tan(x) + \frac{5}{16}\sec(x)\tan(x) + \frac{5}{16}\ln\left|\sec(x)+\tan(x)\right| -$$

$$2\left(\frac{1}{4}\sec^3(x)\tan(x) + \frac{3}{8}\sec(x)\tan(x) + \frac{3}{8}\ln\left|\sec(x)+\tan(x)\right|\right)$$

$$+ \int\sec^3(x)\,dx = \frac{1}{2}\sec(x)\tan(x) + \frac{1}{2}\ln\left|\sec(x)+\tan(x)\right| + C$$

$$\int\tan^4(x)\sec^3(x)\,dx =$$

$$\frac{1}{6}\sec^5(x)\tan(x) - \frac{7}{24}\sec^3(x)\tan(x) + \frac{5}{8}\sec(x)\tan(x) + \frac{5}{8}\ln\left|\sec(x)+\tan(x)\right| + C$$

Example 15: Evaluate $\int\tan^6(2x)\sec^5(2x)\,dx$.

Solution: Beware! There will be a lot of algebra in this problem.

1. Rewrite: $\int\tan^6(2x)\sec^5(2x)\,dx = \int\left(\tan^2(2x)\right)^3\sec^5(2x)\,dx =$

$$\int\left(\sec^2(2x)-1\right)^3\sec^5(2x)\,dx .$$

2. Binomial Expansion: $\left(\sec^2(2x)-1\right)^3 = \sec^6(2x) - 3\sec^4(2x) + 3\sec^2(2x) - 1$.

3. So $\int\left(\sec^2(2x)-1\right)^3\sec^5(2x)\,dx = \int\sec^{11}(2x) - 3\sec^9(2x) + 3\sec^7(2x) - \sec^5(2x)\,dx$:

$$\int\sec^5(2x)\,dx = \frac{1}{2}\left(\frac{1}{4}\sec^3(2x)\tan(2x) + \frac{3}{8}\sec(2x)\tan(2x) + \frac{3}{8}\ln\left|\sec(2x)+\tan(2x)\right|\right)$$

$$= \frac{1}{8}\sec^3(2x)\tan(2x) + \frac{3}{16}\sec(2x)\tan(2x) + \frac{3}{16}\ln\left|\sec(2x)+\tan(2x)\right|$$

$$\int\sec^7(2x)\,dx = \frac{1}{2}\left(\frac{1}{6}\sec^5(2x)\tan(2x) + \frac{5}{6}\int\sec^5(2x)\,dx\right)$$

$$= \frac{1}{12}\sec^5(2x)\tan(2x) + \left(\frac{5}{6}\right)\left(\frac{1}{2}\right)\left(\frac{1}{4}\sec^3(2x)\tan(2x) + \frac{3}{8}\sec(2x)\tan(2x) +\right.$$

$$\left.\frac{3}{8}\ln\left|\sec(2x)+\tan(2x)\right|\right)$$

$$= \frac{1}{12}\sec^5(2x)\tan(2x) + \frac{5}{48}\sec^3(2x)\tan(2x) + \frac{5}{32}\sec(2x)\tan(2x) + \cdot$$

$$\frac{5}{32}\ln\left|\sec(2x)+\tan(2x)\right|$$

$$\int \sec^9(2x)\,dx \;=\; \tfrac{1}{2}\Big(\tfrac{1}{8}\sec^7(2x)\tan(2x) + \tfrac{7}{8}\int \sec^7(2x)\,dx\Big)$$

$$=\; \tfrac{1}{16}\sec^7(2x)\tan(2x) + \tfrac{7}{8}\big(\tfrac{1}{2}\big)\Big(\tfrac{1}{12}\sec^5(2x)\tan(2x) + \tfrac{5}{48}\sec^3(2x)\tan(2x) + $$

$$\tfrac{5}{32}\ln\big|\sec(2x) + \tan(2x)\big|\Big)$$

$$=\; \tfrac{1}{16}\sec^7(2x)\tan(2x) + \tfrac{7}{192}\sec^5(2x)\tan(2x) + \tfrac{35}{768}\sec^3(2x)\tan(2x) + $$

$$\tfrac{35}{512}\ln\big|\sec(2x) + \tan(2x)\big|$$

$$\int \sec^{11}(2x)\,dx \;=\; \tfrac{1}{20}\sec^9(2x)\tan(2x) + \tfrac{9}{20}\int \sec^9(2x)\,dx$$

$$=\; \tfrac{1}{20}\sec^9(2x)\tan(2x) + \big(\tfrac{9}{10}\big)\big(\tfrac{1}{2}\big)\Big(\tfrac{1}{16}\sec^7(2x)\tan(2x) + \tfrac{7}{192}\sec^5(2x)\tan(2x) + $$

$$\tfrac{35}{768}\sec^3(2x)\tan(2x) + \tfrac{35}{512}\sec(2x)\tan(2x) + \tfrac{35}{512}\ln\big|\sec(2x) + \tan(2x)\big|\Big)$$

$$=\; \tfrac{1}{20}\sec^9(2x)\tan(2x) + \tfrac{9}{320}\sec^7(2x)\tan(2x) + \tfrac{21}{1280}\sec^5(2x)\tan(2x) + $$

$$\tfrac{3}{32}\sec^3(2x)\tan(2x) + \tfrac{63}{2048}\sec(2x)\tan(2x) + \tfrac{63}{2048}\ln\big|\sec(2x) + \tan(2x)\big|$$

4. Therefore, $\displaystyle\int \tan^6(2x)\sec^5(2x)\,dx \;=\;$

$$\tfrac{1}{8}\sec^3(2x)\tan(2x) + \tfrac{3}{16}\sec(2x)\tan(2x) + \tfrac{3}{16}\ln\big|\sec(2x) + \tan(2x)\big| :$$

$$+ \;\; \tfrac{1}{12}\sec^5(2x)\tan(2x) + \tfrac{5}{48}\sec^3(2x)\tan(2x) + \tfrac{5}{32}\sec(2x)\tan(2x) + \tfrac{5}{32}\ln\big|\sec(2x) + \tan(2x)\big|$$

$$+ \;\; \tfrac{1}{16}\sec^7(2x)\tan(2x) + \tfrac{7}{192}\sec^5(2x)\tan(2x) + \tfrac{35}{768}\sec^3(2x)\tan(2x) + \tfrac{35}{512}\sec(2x)\tan(2x) + $$

$$\tfrac{35}{512}\ln\big|\sec(2x) + \tan(2x)\big|$$

$$+ \;\; \tfrac{1}{20}\sec^9(2x)\tan(2x) + \tfrac{9}{320}\sec^7(2x)\tan(2x) + \tfrac{21}{1280}\sec^5(2x)\tan(2x) + \tfrac{3}{32}\sec^3(2x)\tan(2x) + $$

$$\tfrac{63}{2048}\sec(2x)\tan(2x) + \tfrac{63}{2048}\ln\big|\sec(2x) + \tan(2x)\big|$$

$$\int \tan^6(2x)\sec^5(2x)\,dx \;=\; \tfrac{1}{20}\sec^9(2x)\tan(2x) + \tfrac{29}{320}\sec^7(2x)\tan(2x) + \tfrac{523}{3840}\sec^5(2x)\tan(2x)$$

$$+ \;\; \tfrac{283}{768}\sec^3(2x)\tan(2x) + \tfrac{907}{2048}\sec(2x)\tan(2x) + \tfrac{907}{2048}\ln\big|\sec(2x) + \tan(2x)\big| + C$$

Understand that there is no way anybody will ever ask you to repeat this without some type of electronic integration tool available to you.

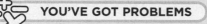

YOU'VE GOT PROBLEMS

Problem 6: Evaluate $\int \tan^2(x)\sec^3(x)\,dx$.

The Least You Need to Know

- Use right triangle trigonometry when evaluating $\int \sqrt{x^2 - a^2}\,dx, \int \sqrt{x^2 + a^2}\,dx, \int \sqrt{a^2 - x^2}\,dx$.
- Use the double angle identity for cosine when the integrand is $\sin^2(x)$ or $\cos^2(x)$.
- Use the Pythagorean identity $1 + \tan^2(x) = \sec^2(x)$ when working with the product of powers of the tangent and secant functions.

CHAPTER 8

Integration with Fractions

This is the last of the chapters on integration techniques. (Although Chapter 9 does deal with integration and issues with infinity.) In this chapter, we'll study how to work with integrands that take particular fractional forms and transform them into manageable integrands. This would be a good time for you to look back at Chapter 1 and the process of dismantling fractional expressions.

We'll finish the chapter with a number of integration problems that incorporate all we've studied in this book up to this point.

In This Chapter

- Completing the square
- Integration by partial fractions
- Working with nonrepeated linear factors
- Repeated linear factors
- Understanding irreducible quadratic factors

Completing the Square

Before we do some calculus, let's take a minute to review a little bit of algebra. The expression $\left(a+b\right)^2$ is clearly a square. The exponential 2 gives it away. When you expand $\left(a+b\right)^2$ you get $a^2+2ab+b^2$. This is called a square trinomial. Get it? It's a square and it has three terms. Observe that the first and third terms are squared, and the middle term is twice the product of the individual terms of the original binomial. The expression $(2x+3)^2 = (2x)^2 + 2(2x)(3) + (3)^2 = 4x^2 + 12x + 9$. The expression $(a-b)^2 = a^2 - 2ab + b^2$ is also a square trinomial.

The trinomial $x^2 + 8x + 9$ is not a square trinomial. Although the first and third terms are squares, the middle term is not twice the product of x and 3. There is a technique called completing the square that is useful in algebra for creating equations in standard forms and in calculus for transforming an integrand into a recognizable integration form. See the upcoming examples to learn how to complete the square.

✏️ **CRITICAL POINT**

To complete the square on a quadratic $ax^2 + bx + c$: Factor the leading coefficient from the first two terms of the quadratic to create the expression $a\left(x^2 + \frac{b}{a}x\right) + c$. Halve the linear term, $\frac{b}{2a}$, square it, add it within the parentheses, and subtract a times this value from c:

$a\left(x^2 + \frac{b}{a}x + \left(\frac{b}{2a}\right)^2\right) + c - a\left(\frac{b}{2a}\right)^2$. The expression is now $a\left(x + \frac{b}{a}\right)^2 + c - \frac{b^2}{4a} = a\left(x + \frac{b}{a}\right)^2 + \frac{4ac-b^2}{4a}$.

Example 1: Complete the square with $4x^2 + 12x + 10$.

Solution:

1. Factor out the 4, $4(x^2 + 3x) + 10$.

2. Take half of 3, square it, and add the value inside the parentheses, while subtracting 4 times this amount from the 9:

$$4\left(x^2 + 3x + \left(\tfrac{3}{2}\right)^2\right) + 10 - 4\left(\tfrac{3}{2}\right)^2$$

3. Simplify this expression by writing the trinomial as a squared binomial and performing the arithmetic at the end of the expression:

$$4\left(x + \tfrac{3}{2}\right)^2 + 10 - 9 = 4\left(x + \tfrac{3}{2}\right)^2 + 1$$

Example 2: Evaluate $\int \frac{1}{4x^2+12x+10}dx$.

Solution:

1. The denominator is the same as the trinomial from Example 1.

2. Completing the square on the denominator gives $\int \frac{1}{4\left(x+\frac{3}{2}\right)^2+1}dx$.

3. The denominator is the sum of two squares and can fit the form for the integral problem $\int \frac{1}{1+u^2}du = \tan^{-1}(u) + C.$

4. The question is what do we do about that 4?

5. Rewrite $4\left(x+\frac{3}{2}\right)^2$ as $\left(2\left(x+\frac{3}{2}\right)\right)^2 = \left(2x+3\right)^2$.

6. Let $u = 2x + 3$, which makes $du = 2dx$ or $dx = \frac{1}{2}du$.

7. Transform the integral to $\frac{1}{2}\int \frac{1}{1+u^2}du = \frac{1}{2}\tan^{-1}(u) + C$. Transforming back to the original variable, $\int \frac{1}{4x^2+12x+10}dx = \frac{1}{2}\tan^{-1}(2x+3) + C.$

Example 3: Evaluate $\int \frac{1}{3x^2+6x+4}dx$.

Solution:

1. Complete the square in the denominator: $3x^2 + 6x + 4 = 3(x^2 + 2x) + 4 = 3(x^2 + 2x + 1) + 4 - 3.$

2. Simplify this expression to be $3(x + 1)^2 + 1$. The integrand is now $\int \frac{1}{3(x+1)^2+1}dx$.

3. In order to perform the u-substitution, we need to move the 3 inside the parentheses. This is not going to be as "clean" as moving the 4 inside the parentheses in Example 2:

   $$3(x + 1)^2 + 1 = \left(\sqrt{3}\left(x+1\right)\right)^2 + 1$$

4. Let $u = \sqrt{3}\left(x+1\right)$, so $du = \sqrt{3}\,dx$ and $dx = \frac{1}{\sqrt{3}}\,du$.

5. Transform the integral to $\frac{1}{\sqrt{3}}\int \frac{1}{u^2+1}du = \frac{1}{\sqrt{3}}\tan^{-1}(u) + C$.

6. Working back to the original variable, $\int \frac{1}{3x^2+6x+4}dx = \frac{1}{\sqrt{3}}\tan^{-1}\left(\sqrt{3}\left(x+1\right)\right) + C$.

Example 4: Evaluate $\int \frac{1}{x^2+8x+20} dx$.

Solution:

1. Complete the square in the denominator: $x^2 + 8x + 20 = (x^2 + 8x + 16) + 4 = (x + 4)^2 + 4$.

2. Rewrite the integrand: $\int \frac{1}{(x+4)^2+4} dx$. The form of the integrand for $\tan^{-1}(u)$ is of the form $\frac{1}{1+u^2}$, not $\frac{1}{4+u^2}$. We need to factor out the 4 as follows:

$$\int \frac{1}{(x+4)^2+4} dx = \frac{1}{4}\int \frac{1}{\frac{(x+4)^2}{4}+1} dx = \frac{1}{4}\int \frac{1}{\left(\frac{x+4}{2}\right)^2+1} dx$$

3. Let $u = \frac{x+4}{2}$ so that $du = \frac{1}{2} dx$ or that $dx = 2du$:

$$\int \frac{1}{\left(\frac{x+4}{2}\right)^2+1} dx = \frac{1}{2}\int \frac{1}{1+u^2} du = \frac{1}{2} \tan^{-1}(u) + C$$

4. Transform the result back to the original variable, $\int \frac{1}{x^2+8x+20} dx = \frac{1}{2}\tan^{-1}\left(\frac{x+4}{2}\right)+C$.

YOU'VE GOT PROBLEMS

Problem 1: Evaluate $\int \frac{1}{9x^2+12x+20} dx$.

The integration formula for inverse tangent is not the only situation in which completing the square will be useful. The formula for inverse sine will also be helpful. You'll need to be careful with the arithmetic because the radicand in the denominator is $1 - u^2$ and that nasty subtraction symbol, like a spoiled child, causes a lot of trouble when it isn't given enough attention.

Example 5: Evaluate $\int \frac{1}{\sqrt{13-12x-x^2}} dx$.

Solution:

1. Complete the square within the radicand while paying attention to how we deal with the subtraction:

$$\sqrt{13-12x-x^2} = \sqrt{13-\left(x^2+12x\right)} = \sqrt{13-\left(x^2+12x+36\right)+36}$$

2. Do you see why the 36 was added at the end rather than subtracted? The 36 inside the parentheses is affected by the subtraction sign appearing before the parentheses. That means we need to add 36 to keep the balance:

$$\sqrt{13-\left(x^2+12x+36\right)+36} = \sqrt{49-\left(x+6\right)^2}$$

3. Factor 49 from the radicand so that it takes the form $1 - u^2$:

$$\sqrt{49 - (x + 6)^2} = \sqrt{49}\sqrt{1 - \frac{(x+6)^2}{49}} = 7\sqrt{1 - \left(\frac{x+6}{7}\right)^2}$$

4. Therefore, $\int \frac{1}{\sqrt{13 - 12x - x^2}} dx = \int \frac{1}{7\sqrt{1 - \left(\frac{x+6}{7}\right)^2}} dx = \frac{1}{7}\int \frac{1}{\sqrt{1 - \left(\frac{x+6}{7}\right)^2}} dx$. Let $u = \frac{x+6}{7}$ do that

   $du = \frac{1}{7} dx$:

$$\frac{1}{7}\int \frac{1}{\sqrt{1 - \left(\frac{x+6}{7}\right)^2}} dx = \int \frac{1}{\sqrt{1 - u^2}} du = \sin^{-1}(u) + C$$

5. Return to the original variable:

$$\int \frac{1}{\sqrt{13 - 12x - x^2}} dx = \sin^{-1}\left(\frac{x+6}{7}\right) + C$$

Example 6: Evaluate $\int \frac{1}{\sqrt{1 - 2x - 2x^2}} dx$.

Solution:

1. Complete the square for $1 - 2x - 2x^2$: $1 - 2(x^2 + x)$ becomes $1 - 2\left(x^2 + x + \left(\frac{1}{2}\right)^2\right) + \frac{1}{2} = \frac{3}{2} - 2\left(x + \frac{1}{2}\right)^2$.

2. The radicand is supposed to be of the form $1 - u^2$, so we'll need to factor out $\frac{3}{2}$:

$$\frac{3}{2} - 2\left(x + \frac{1}{2}\right)^2 = \frac{3}{2}\left(1 - \frac{4}{3}\left(x + \frac{1}{2}\right)^2\right).$$

3. Now we'll have to move the $\frac{4}{3}$ inside the parentheses:

$$\frac{3}{2}\left(1 - \frac{4}{3}\left(x + \frac{1}{2}\right)^2\right) = \frac{3}{2}\left(1 - \left(\frac{2}{\sqrt{3}}x + \frac{1}{\sqrt{3}}\right)^2\right)$$

4. Let's see what we can do with this. $\int \frac{1}{\sqrt{1 - 2x - 2x^2}} dx$ becomes $\int \frac{1}{\sqrt{\frac{3}{2}\left(1 - \left(\frac{2}{\sqrt{3}}x + \frac{1}{\sqrt{3}}\right)^2\right)}} dx$ and this

   equals $\sqrt{\frac{2}{3}}\int \frac{1}{\sqrt{1 - \left(\frac{2}{\sqrt{3}}x + \frac{1}{\sqrt{3}}\right)^2}} dx$.

5. Let $u = \frac{2}{\sqrt{3}}x + \frac{1}{\sqrt{3}}$ so that $du = \frac{2}{\sqrt{3}} dx$ and $\frac{\sqrt{3}}{2} du = dx$.

6. The integral transforms to $\left(\frac{\sqrt{3}}{2}\right)\left(\sqrt{\frac{2}{3}}\right)\int \frac{1}{\sqrt{1 - u^2}} du = \frac{\sqrt{2}}{2}\sin^{-1}(u) + C$, bringing this back to

   the original variable, $\int \frac{1}{\sqrt{1 - 2x - 2x^2}} dx = \frac{\sqrt{2}}{2}\sin^{-1}\left(\frac{2}{\sqrt{3}}x + \frac{1}{\sqrt{3}}\right) + C$.

YOU'VE GOT PROBLEMS

Problem 2: Evaluate $\int \frac{1}{\sqrt{-4x^2 - 6x - 2}} dx$.

Integration by Partial Fractions

Any number of fractional factors could make up the partial fractions. We'll only consider three cases:

- Nonrepeating linear factors
- Repeating linear factors
- Irreducible quadratic factors

Nonrepeating Linear Factors

Example 7: Evaluate $\int \frac{1}{x^2-9} dx$.

Solution:

1. As you know, $x^2 - 9 = (x + 3)(x - 3)$, so we will write the equation $\frac{1}{x^2-9} = \frac{A}{x+3} + \frac{B}{x-3}$.

2. Multiply both sides of the equation by $(x + 3)(x - 3)$ to get $1 = A(x - 3) + B(x + 3)$.

3. Set $x = 3$, thus eliminating the first part of the right-hand side of the equation, to get
    $1 = 6B$ so $B = \frac{1}{6}$.

4. Now set $x = -3$, eliminating the second part of the right side of the equation, to get
    $1 = -6A$ or $A = \frac{-1}{6}$.

5. Rewrite $\int \frac{1}{x^2-9} dx$ as $\int \frac{1}{6}\left(\frac{1}{x-3} - \frac{1}{x+3}\right)dx = \frac{1}{6}\int\left(\frac{1}{x-3} - \frac{1}{x+3}\right)dx = \frac{1}{6}\left(\ln|x-3| - \ln|x+3|\right) + C$
    or $\frac{1}{6}\ln\left|\frac{x-3}{x+3}\right| + C$.

✏️ **CRITICAL POINT**

We are looking at cases where the derivative of the denominator is not present in the problem. Given that, when do you use partial fractions versus when do you complete the square? If you can factor the denominator, use partial fractions. If not, go to the process of completing of the square.

Example 8: Evaluate $\int \frac{1}{4x^2+13x+3}dx$.

Solution:

1. Rewrite the denominator as $(4x + 1)(x + 3)$.

2. Solve for the coefficients using partial fractions.

   $\frac{1}{4x^2+13x+3} = \frac{A}{4x+1} + \frac{B}{x+3}$ becomes $1 = A(x + 3) + B(4x + 1)$

3. Set $x = -3$ to get $1 = -11B$ or $B = \frac{-1}{11}$.

4. Set $x = \frac{-1}{4}$ to get $1 = \frac{11}{4}A$ so that $A = \frac{4}{11}$.

5. We now have $\int \frac{1}{4x^2+13x+3}dx = \int \frac{4}{11}\left(\frac{1}{4x+1}\right) - \frac{1}{11}\left(\frac{1}{x+3}\right)dx = \frac{1}{11}\int \frac{4}{4x+1} - \frac{1}{x+3}dx =$
   $\frac{1}{11}\left(\ln|4x+1| - \ln|x+3|\right) + C = \frac{1}{11}\ln\left|\frac{4x+1}{x+3}\right| + C$.

Example 9: Evaluate $\int \frac{1}{4x^3-9x}dx$.

Solution:

1. The denominator factors to $x(2x + 3)(2x - 3)$.

2. Create the equation for decomposition:

   $\frac{1}{4x^3-9x} = \frac{A}{x} + \frac{B}{2x+3} + \frac{C}{2x-3}$

3. Multiply both sides of the equation by $x(2x + 3)(2x - 3)$:
   $1 = A(2x + 3)(2x - 3) + Bx(2x - 3) + Cx(2x + 3)$

4. Set $x = 0$: $1 = -9A$ so $A = \frac{-1}{9}$.

5. Set $x = \frac{-3}{2}$: $1 = B\left(\frac{-3}{2}\right)(-6) = 9B$ so $B = \frac{1}{9}$.

6. Set $x = \frac{3}{2}$: $1 = C\left(\frac{3}{2}\right)(6) = 9C$ so $C = \frac{1}{9}$.

7. $\int \frac{1}{4x^3-9x}dx$ becomes $\frac{1}{9}\int \frac{1}{2x+3} + \frac{1}{2x-3} - \frac{1}{x}dx = \frac{1}{9}\left(\frac{1}{2}\ln|2x+3| + \frac{1}{2}\ln|2x-3| - \ln|x|\right) + C =$
   $\frac{1}{9}\ln\left|\frac{\sqrt{(2x+3)(2x-3)}}{x}\right| + C$.

YOU'VE GOT PROBLEMS

Problem 3: Evaluate $\int \frac{1}{3x^2+10x+8}dx$.

Repeated Linear Factors

Integrating $\int \frac{1}{(x+2)^2} dx$ is straightforward because you can use u-substitution with $u = x + 2$ and arrive at the answer $\frac{-1}{x+2} + C$. In this section, we'll examine rational integrands with three or more factors, with one factor being repeated.

✏️ **CRITICAL POINT**

When one or more linear factors are repeated in the denominator of an integrand, it is necessary to include one more fraction on the right side of the equation when decomposing the integrand into partial fractions.

Example 10: Evaluate $\int \frac{x^2}{(x+1)^3} dx$.

Solution:

1. First, take note the x^2 is not the derivative, or a multiple of the derivative, of $(x + 1)^3$.

2. When we write the equation to decompose the integrand into partial fractions, we need to realize that $x + 1$ is a factor three times.

3. Therefore, we'll need a fraction for $x + 1$, a fraction for $(x + 1)^2$, and a fraction for $(x + 1)^3$. Each is a linear factor, so the numerators in each case will be a constant.

4. Create the equation for decomposition:

   $$\frac{x^2}{(x+1)^3} = \frac{A}{x+1} + \frac{B}{(x+1)^2} + \frac{C}{(x+1)^3}$$

5. Multiply both sides of the equation by the common denominator $(x + 1)^3$:

   $$x^2 = A(x + 1)^2 + B(x + 1) + C$$

6. The value of C is easy to determine. Set $x = -1$: $(-1)^2 = A(0) + B(0) + C$ so $C = 1$.

7. We will not be able to eliminate other coefficients as we try to determine the values of A and B. We will need to create a system of equations.

8. Set $x = 1$ (because 1 is an easy number): $1 = 4A + 2B + 1$ so $4A + 2B = 0$.

9. Set $x = 2$ (another easy number): $2^2 = 9A + 3B + 1$ so $9A + 3B = 3$.

10. Solve the system of equations to determine that $A = 1$ and $B = -2$.

11. Rewrite the integral into the decomposed fractions (no, they are not zombies):

    $$\int \frac{x^2}{(x+1)^3} dx = \int \frac{1}{x+1} dx - 2\int \frac{1}{(x+1)^2} dx + \int \frac{1}{(x+1)^3} dx = \ln|x+1| + \frac{2}{(x+1)} - \frac{1}{2(x+1)^2} + C$$

Example 11: $\int \frac{2x+5}{4x^3+4x^2+x} dx$.

Solution:

1. The denominator factors to be $x(2x + 1)^2$. The equation to decompose the fraction is:

    $$\frac{2x+5}{4x^3+4x^2+x} = \frac{A}{x} + \frac{B}{2x+1} + \frac{C}{(2x+1)^2}$$

2. Multiply both sides of the equation by the common denominator to get:

    $2x + 5 = A(2x + 1)^2 + Bx(2x + 1) + Cx$

3. Set $x = 0$: $5 = A$.

4. Set $x = \left(\frac{-1}{2}\right)$: $2\left(\frac{-1}{2}\right) + 5 = C\left(\frac{-1}{2}\right)$ so that $C = -8$.

5. Let $x = 1$ (remember, an easy number): $2(1) + 3 = 5(2(1)+1)^2 + B(1)(3) - 8$.

    $7 = 45 + 3B - 8$ so that $3B = -30$ or $B = -10$

6. Rewrite the integral into the decomposed fractions:

    $$\int \frac{2x+5}{4x^3+4x^2+x} dx = \int \frac{5}{x} - 10\left(\frac{1}{2x+1}\right) - \frac{8}{(2x+1)^2} dx = 5\ln|x| - 5\ln|2x+1| + \frac{4}{2x+1} + C$$

YOU'VE GOT PROBLEMS

Problem 4: Evaluate $\int \frac{2x+3}{(x+1)(3x-2)^2} dx$.

Irreducible Quadratic Factors

The third fractional factor that could make up a partial fraction is quadratics that cannot be factored with the real numbers. Since the numerator of the factors has a degree 1 less than the denominator, the numerators for the quadratic factors will be linear rather than constants.

Example 12: Evaluate $\int \frac{1}{x^3+x} dx$.

Solution:

1. The denominator factors to $x(x^2 + 1)$. The equation to decompose the fraction is:

    $$\frac{1}{x^3+x} = \frac{A}{x} + \frac{Bx+C}{x^2+1}$$

2. Multiply through by the common denominator to get:

    $1 = A(x^2 + 1) + (Bx + C)(x)$

3. Set $x = 0$: $1 = A + 0$ so $A = 1$.

4. Set $x = -1$: $1 = 2 + B - C$.

5. Set $x = 1$ (easy number): $1 = 2 + B + C$. Solving the system, givens $B = -1$ and $C = 0$.

6. Rewrite the integral into the decomposed fractions:

$$\int \frac{1}{x^3+x}\,dx = \int \frac{1}{x} - \frac{x}{x^2+1}\,dx = \ln|x| - \frac{1}{2}\ln|x^2+1| + C$$

Example 13: Evaluate $\int \frac{x}{x^3-1}\,dx$.

Solution:

1. Recall that $x^3 - 1 = (x-1)(x^2+x+1)$. The equation to decompose the fraction is:

    $$\frac{x}{x^3-1} = \frac{A}{x-1} + \frac{Bx+C}{x^2+x+1}$$

2. Multiply through by the common denominator:

    $$x = A(x^2+x+1) + (Bx+C)(x-1)$$

3. Set $x = 1$: $1 = A(1+1+1) + 0$ so $A = \frac{1}{3}$.

4. Set $x = 0$: $0 = (\frac{1}{3})(1) + C(-1)$ so $C = \frac{1}{3}$.

5. Set $x = -1$ (easy number): $-1 = (\frac{1}{3})(1) + (-B + \frac{1}{3})(-2)$ so $-2(-B + \frac{1}{3}) = \frac{-4}{3}$, and $B = \frac{-1}{3}$.

6. Rewrite the integral into the decomposed fractions:

    $$\int \frac{x}{x^3-1}\,dx = \frac{1}{3}\int \frac{1}{x-1} + \frac{-x+1}{x^2+x+1}\,dx$$

7. The antiderivative of $\frac{1}{x-1}$ is $\ln|x-1|$. That is not a problem. It's the second part of the integrand that requires us to be careful.

8. The derivative of $x^2 + x + 1$ is $2x + 1$.

9. We don't have $2x$, we have x. Half of $2x + 1$ is $x + \frac{1}{2}$, so we can rewrite $\frac{-x+1}{x^2+x+1}$ as
    $$\frac{-(x+\frac{1}{2})}{x^2+x+1} + \frac{\frac{3}{2}}{x^2+x+1}.$$

10. The antiderivative of $\frac{-(x+\frac{1}{2})}{x^2+x+1}$ is $-\frac{1}{2}\ln|x^2+x+1|$.

11. This leaves us with $\frac{\frac{3}{2}}{x^2+x+1}$.

12. We have a quadratic denominator that cannot be factored with a constant numerator. Time to complete the square in the denominator.

$$x^2 + x + 1 = \left(x^2 + x + \left(\tfrac{1}{2}\right)^2\right) + 1 - \left(\tfrac{1}{2}\right)^2 = \left(x + \tfrac{1}{2}\right)^2 + \tfrac{3}{4}$$

13. Remember, we're trying to match this denominator to the inverse tangent form $1 + u^2$.

    We need to rewrite $\left(x + \tfrac{1}{2}\right)^2 + \tfrac{3}{4}$ as $\tfrac{3}{4}\left(\tfrac{4}{3}\left(x + \tfrac{1}{2}\right)^2 + 1\right)$, which equals $\tfrac{3}{4}\left(\left(\tfrac{2x+1}{\sqrt{3}}\right)^2 + 1\right)$.

14. The antiderivative of $\dfrac{\frac{3}{2}}{\frac{3}{4}\left(\left(\frac{2x+1}{\sqrt{3}}\right)^2 + 1\right)} = \dfrac{2}{\left(\left(\frac{2x+1}{\sqrt{3}}\right)^2 + 1\right)}$ is $\sqrt{3}\tan^{-1}\left(\tfrac{2x+1}{\sqrt{3}}\right)$.

15. Therefore, $\displaystyle\int \frac{x}{x^3-1}\,dx = \frac{1}{3}\int \frac{1}{x-1} - \frac{1}{2}\left(\frac{x+\frac{1}{2}}{x^2+x+1}\right) + \frac{2}{\left(\frac{2x+1}{\sqrt{3}}\right)^2 + 1}\,dx =$

$$\frac{1}{3}\left(\ln|x-1| - \frac{1}{2}\ln\left|x^2+x+1\right| + \sqrt{3}\tan^{-1}\left(\tfrac{2x+1}{\sqrt{3}}\right)\right) + C.$$ Wow! That was a mouthful!

YOU'VE GOT PROBLEMS

Problem 5: Evaluate $\displaystyle\int \frac{5}{x^3+4x}\,dx$.

The Least You Need to Know

- When the denominator of an integrand is of the form $ax^2 + bx + c$, look to see if completing the square is an option you can use to compute the integral.
- Identify quadratic factors in the denominator of the integrand that allow you to rewrite the fraction as the sum and difference of the component fractions.
- You can find the appropriate numerators for the component fractions based on the pattern of the denominator's factors.

The Infinite Series and More

We begin this part of the book by computing limits with infinite bounds. Then we compute limits with finite bounds, but with functions that have infinite discontinuities so we can cover our bases on all types of integration issues. We then change our focus from the traditional $y = f(x)$ to examine functions defined parametrically and functions defined in the polar coordinate plane.

An introduction to Vector Calculus and Differential Equations finishes the preparatory work needed before we look at the ever-important topics of sequences and series. Power series, including the special cases of MacLaurin and Taylor Series, give us an eye-opener as to how our calculators work.

Finally, the last chapter in the book is a final exam covering all you've learned throughout the book. Good luck!

To Infinity and Beyond

We looked at the Fundamental Theorem of Calculus, and we have concentrated on studying techniques of integration. In this chapter, we will go back to definite integrals and look at problems that either have an infinite bound of integration or an infinite discontinuity in a finite region.

In This Chapter

* Exploring improper integrals
* Working with infinite limits of integration
* Examining discontinuities in the integrand
* Using the Comparison Test for improper integrals

Improper Integrals

In all cases involving *improper integrals*, we will treat the problem as a limit. If the limit exists, we say that the integral *converges* and if the limit does not exist, we say that the integral *diverges*.

> 📖 **DEFINITION**
>
> An **improper integral** is a definite integral that has either or both bounds of integration going to an infinity, or integrands that approach infinity at one or more points in the range of integration.

Infinite Limits of Integration

Consider the area under the graph of $f(x) = \frac{1}{x^2}$ and above the x-axis on the interval $[1, n]$. The integral $\int_1^n \frac{1}{x^2}\,dx$ computes the amount of area.

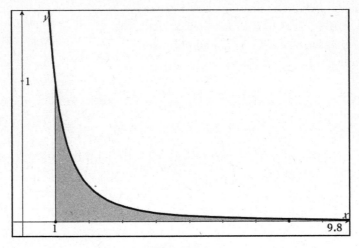

Figure 9.1

The area under the curve $y = \frac{1}{x^2}$ from $x = 1$ to infinity.

$$\int_1^n \frac{1}{x^2}\,dx = \frac{-1}{x}\Big|_1^n = \frac{-1}{n} - \frac{-1}{1} = 1 - \frac{1}{n}$$

What happens to this area as n gets very large? The larger the value of n, the closer $\frac{1}{n}$ gets to 0 and the closer $1 - \frac{1}{n}$ gets to 1. That is, $\lim\limits_{n \to \infty} \int_1^n \frac{1}{x^2}\,dx = 1$.

In this case, we can state that $\int_1^\infty \frac{1}{x^2}\,dx = 1$.

Do we get the same result for $f(x) = \frac{1}{x}$: $\int_1^\infty \frac{1}{x}\,dx = \lim\limits_{n \to \infty} \int_1^n \frac{1}{x}\,dx = \lim\limits_{n \to \infty}\left(\ln(x)\Big|_1^n\right) = \lim\limits_{n \to \infty}\left(\ln(n) - \ln(1)\right) = \lim\limits_{n \to \infty}\left(\ln(n)\right)$.

As n gets extremely large, $\ln(n)$ also get extremely large so $\lim\limits_{n \to \infty}\left(\ln(n)\right)$ fails to exist, and the integral is said to be divergent.

CRITICAL POINT

If the integral of f(x) on [a, n] exists for all values of $n > a$, then $\int_a^\infty f(x)\,dx = \lim\limits_{n \to \infty} \int_a^n f(x)\,dx$ provided the limit exists and is finite.

We say the integrals are convergent if the limits exist. We say that the integrals are divergent if the limits fail to exist.

Example 1: For what values of p is the integral $\int_a^\infty \frac{1}{x^p}\,dx$ convergent?

Solution:

1. $\int_a^\infty \frac{1}{x^p}\,dx = \lim\limits_{n \to \infty} \int_a^n \frac{1}{x^p}\,dx = \lim\limits_{n \to \infty}\left(\left(\frac{1}{-p+1}\right)\frac{1}{x^{p-1}}\Big|_a^n\right) = \frac{1}{1-p}\left(\lim\limits_{n \to \infty}\left(\frac{1}{n^{p-1}} - \frac{1}{a^{p-1}}\right)\right)$.

2. If $p > 1$, $\frac{1}{1-p}\left(\lim\limits_{n \to \infty}\left(\frac{1}{n^{p-1}}\right)\right) = 0$ while if $p \leq 1$ and not equal to 0, $\frac{1}{1-p}\left(\lim\limits_{n \to \infty}\left(\frac{1}{n^{p-1}}\right)\right)$ diverges.

3. Therefore, $\int_a^\infty \frac{1}{x^p}\,dx$ is convergent for $p > 1$.

CRITICAL POINT

The integral $\int_a^\infty \frac{1}{x^p}\,dx$ will converge whenever $p > 1$ and will diverge when $p \leq 1$. This is called the p-Test for improper integrals.

Example 2: Evaluate $\int_{-\infty}^{0} e^x \, dx$, if it exists.

Solution: $\lim_{n \to -\infty} \left(\int_{n}^{0} e^x \, dx \right) = \lim_{n \to -\infty} (e^0 - e^n) = 1 - 0 = 1$.

Example 3: Evaluate $\int_{-\infty}^{\infty} \frac{1}{x^2 + 1} \, dx$, if it exists.

Solution:

1. Let's treat this as two problems: $\int_{-\infty}^{0} \frac{1}{x^2 + 1} \, dx$ and $\int_{0}^{\infty} \frac{1}{x^2 + 1} \, dx$.

   $\int_{-\infty}^{0} \frac{1}{x^2+1} \, dx = \lim_{n \to -\infty} \left(\int_{n}^{0} \frac{1}{x^2+1} \, dx \right) = \lim_{n \to -\infty} \left(\tan^{-1}(x) \Big|_{n}^{0} \right) = \lim_{n \to -\infty} \left(\tan^{-1}(0) - \tan^{-1}(n) \right) = 0 - \frac{-\pi}{2} = \frac{\pi}{2}$

   In a similar manner:

   $\int_{0}^{\infty} \frac{1}{x^2+1} \, dx = \lim_{n \to \infty} \left(\int_{0}^{n} \frac{1}{x^2+1} \, dx \right) = \lim_{n \to \infty} \left(\tan^{-1}(x) \Big|_{0}^{n} \right) = \lim_{n \to \infty} \left(\tan^{-1}(n) - \tan^{-1}(0) \right) = \frac{\pi}{2} - 0 = \frac{\pi}{2}$

2. Therefore: $\int_{-\infty}^{\infty} \frac{1}{x^2+1} \, dx = \int_{-\infty}^{0} \frac{1}{x^2+1} \, dx + \int_{0}^{\infty} \frac{1}{x^2+1} \, dx = \frac{\pi}{2} + \frac{\pi}{2} = \pi$.

Example 4: Evaluate $\int_{1}^{\infty} (x - 4)e^{-x} \, dx$ if it exists.

Solution:

1. We'll need use integration by parts to evaluate $\int (x - 4)e^{-x} \, dx$. Once we've done that, we can work to the bounds of integration.

   $\int (x-4)e^{-x} \, dx$: Let $u = x - 4$ and $dv = e^{-x} \, dx$. The $du = dx$ and $v = -e^{-x}$.

   $\int (x-4)e^{-x} \, dx = -(x-4)e^{-x} - \int -e^{-x} \, dx = -(x-4)e^{-x} - e^{-x}$.

2. Applying the bounds of integration, you get $\int_{1}^{\infty} (x-4)e^{-x} \, dx = \lim_{n \to \infty} \left(\int_{1}^{n} (x-4)e^{-x} \, dx \right) =$

   $\lim_{n \to \infty} \left(-(x-4)e^{-x} - e^{-x} \Big|_{1}^{n} \right) = \lim_{n \to \infty} \left(\left(-(n-4)e^{-n} - e^{-n} \right) - \left(3e^{-1} - e^{-1} \right) \right) =$

   $\lim_{n \to \infty} \left(\frac{4-n}{e^n} - \frac{1}{e^n} - \frac{2}{e} \right)$.

3. Use L'Hopital's Rule to evaluate $\lim_{n \to \infty} \left(\frac{4-n}{e^n} \right)$. $\lim_{n \to \infty} \left(\frac{4-n}{e^n} \right) = \lim_{n \to \infty} \left(\frac{-1}{e^n} \right) = 0$.

4. Therefore, $\lim_{n \to \infty} \left(\frac{4-n}{e^n} - \frac{1}{e^n} - \frac{2}{e} \right) = \frac{-2}{e}$ and $\int_{1}^{\infty} (x-4)e^{-x} \, dx = \frac{-2}{e}$.

Example 5: Consider the graph of the function h given by $h(x) = e^{-x^2}$ for $0 \leq x < \infty$. Let R be the unbounded region in the first quadrant below the graph of h. Find the volume of the solid generated when R is revolved about the y-axis.

Solution:

1. Use the cylindrical shell method to find the volume (since the region is being rotated about a vertical line).

2. The volume of the solid is $2\pi \int_0^{\infty} xe^{-x^2} dx$, which becomes $2\pi \lim_{n\to\infty} \int_0^n xe^{-x^2} dx$.

   Let $u = -x^2$ so that $du = -2x\, dx$ or $\frac{-1}{2} du = x\, dx$. $\frac{-1}{2}\int e^u\, du = \frac{-1}{2} e^u$.

3. Therefore, $2\pi \int_0^{\infty} xe^{-x^2} dx = -\pi \lim_{n\to\infty}\left(e^{-x^2} \Big|_0^n \right) = -\pi \lim_{n\to\infty}\left(e^{-n^2} - e^0 \right) = -\pi(0-1) = \pi$.

Discontinuities in the Integrand

We've examined what happens when one or both bounds of integration go to infinity. We now consider how to handle discontinuities within the integrand. There are two cases that we should consider:

- The discontinuity is finite.

- The discontinuity is infinite.

CRITICAL POINT

If $f(x)$ is continuous on $[a, b)$ and discontinuous at $x = b$, $\int_a^b f(x)\, dx = \lim_{n\to b^-}\left(\int_a^n f(x)\, dx \right)$ if the limit exists and is finite. Similarly, If $f(x)$ is continuous on $(a, b]$ and discontinuous at $x = a$, $\int_a^b f(x)\, dx = \lim_{n\to a^+}\left(\int_n^b f(x)\, dx \right)$ if the limit exists and is finite.

Example 6: Let $f(x) = \begin{cases} x^2 & x < 3 \\ 2x+5 & x \geq 3 \end{cases}$. Evaluate $\int_0^5 f(x)\,dx$ if it exists.

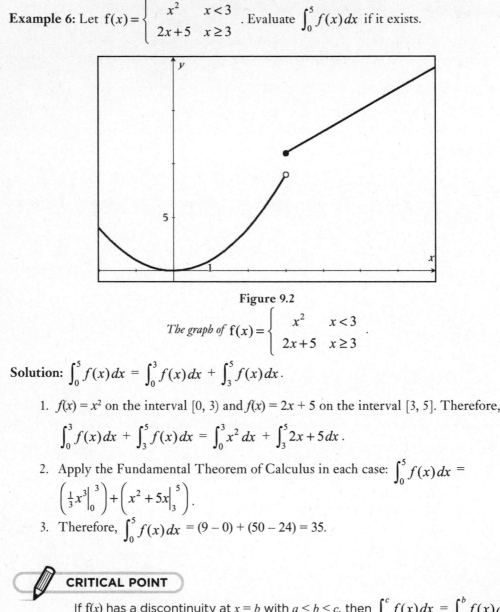

Figure 9.2

The graph of $f(x) = \begin{cases} x^2 & x < 3 \\ 2x+5 & x \geq 3 \end{cases}$.

Solution: $\int_0^5 f(x)\,dx = \int_0^3 f(x)\,dx + \int_3^5 f(x)\,dx$.

1. $f(x) = x^2$ on the interval $[0, 3)$ and $f(x) = 2x + 5$ on the interval $[3, 5]$. Therefore,

$$\int_0^3 f(x)\,dx + \int_3^5 f(x)\,dx = \int_0^3 x^2\,dx + \int_3^5 2x + 5\,dx.$$

2. Apply the Fundamental Theorem of Calculus in each case: $\int_0^5 f(x)\,dx = $
$$\left(\tfrac{1}{3}x^3 \Big|_0^3\right) + \left(x^2 + 5x \Big|_3^5\right).$$

3. Therefore, $\int_0^5 f(x)\,dx = (9 - 0) + (50 - 24) = 35$.

✏️ **CRITICAL POINT**

If f(x) has a discontinuity at $x = b$ with $a < b < c$, then $\int_a^c f(x)\,dx = \int_a^b f(x)\,dx$ $+ \int_b^c f(x)\,dx$ is convergent if both $\int_a^c f(x)\,dx$ and $\int_a^b f(x)\,dx$ are convergent and is divergent if either $\int_a^c f(x)\,dx$ or $\int_a^b f(x)\,dx$ are divergent.

Example 7: Evaluate $\int_0^3 \frac{1}{(x-3)^2} dx$ if it exists.

Solution:

1. Because the integrand does not exist at $x = 3$, we treat $\int_0^3 \frac{1}{(x-3)^2} dx$ as $\lim_{n \to 3^-} \int_0^n \frac{1}{(x-3)^2} dx$.

2. $\lim_{n \to 3^-} \int_0^n \frac{1}{(x-3)^2} dx = \lim_{n \to 3^-} \left(\frac{-1}{x-3} \Big|_0^n \right) = \lim_{n \to 3^-} \left(\frac{-1}{n-3} + \frac{1}{3} \right)$.

3. As n approaches 3 from the left, $\frac{-1}{n-3}$ gets infinitely large, meaning the limit, and therefore the integral, is divergent.

Example 8: Evaluate $\int_0^4 \frac{1}{\sqrt{16-x^2}} dx$ if it exists.

Solution: Since the integrand does not exist at $x = 4$, we treat $\int_0^4 \frac{1}{\sqrt{16-x^2}} dx$ as $\lim_{n \to 4^-} \int_0^n \frac{1}{\sqrt{16-x^2}} dx$.

$$\lim_{n \to 4^-} \int_0^n \frac{1}{\sqrt{16-x^2}} dx = \lim_{n \to 4^-} \left(\int_0^n \frac{1}{4\sqrt{1-\left(\frac{x}{4}\right)^2}} dx \right) = \lim_{n \to 4^-} \left(\sin^{-1}\left(\frac{x}{4}\right) \Big|_0^n \right) =$$

$$\lim_{n \to 4^-} \left(\sin^{-1}\left(\frac{n}{4}\right) - \sin^{-1}(0) \right) = \frac{\pi}{2}$$

Example 9: Evaluate $\int_0^\pi \tan^2(x) dx$ if it exists.

Solution:

1. As we have done before, substitute $\sec^2(x) - 1$ for $\tan^2(x)$.

2. Both the tangent function and the secant function are discontinuous at $x = \frac{\pi}{2}$.

3. Therefore, $\int_0^\pi \sec^2(x) - 1\, dx$ should be written as $\int_0^\pi \sec^2(x) dx - \int_0^\pi 1\, dx$.

4. We know that $\int_0^\pi 1\, dx = \pi$ and does not present a problem.

5. The question of concern is if $\int_0^\pi \sec^2(x) dx = \int_0^{\pi/2} \sec^2(x) dx + \int_{\pi/2}^\pi \sec^2(x) dx$ is convergent.

6. $\int_0^{\pi/2} \sec^2(x) dx = \lim_{n \to \frac{\pi}{2}^-} \int_0^n \sec^2(x) dx = \lim_{n \to \frac{\pi}{2}^-} \left(\tan(x) \Big|_0^n \right) = \lim_{n \to \frac{\pi}{2}^-} \left(\tan(n) - \tan(0) \right) = \infty.$

    Therefore, $\tan(x) \Big|_0^\pi$ is divergent.

BE AWARE

If we had not checked for the point of discontinuity on the interval $[0, \pi]$, then $\int_0^\pi \sec^2(x) dx = \tan(x) \Big|_0^\pi = \tan(\pi) - \tan(0) = 0$, and this is incorrect.

YOU'VE GOT PROBLEMS

Problem 2: Evaluate $\int_{-5}^{0} \frac{1}{\sqrt{25-x^2}}\,dx$ if it exists.

Comparison Test for Improper Integrals

There will be times—particularly when we study infinite series—that all we want to know is if an expression in convergent or divergent. The *Comparison Test for Improper Integrals* helps us with this problem.

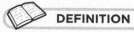 **DEFINITION**

Given f(x) and g(x) are continuous functions for $x > a$ with f(x) $\geq$ g(x) ≥ 0. If $\int_{a}^{\infty} f(x)\,dx$ converges, then so does $\int_{a}^{\infty} g(x)\,dx$. If $\int_{a}^{\infty} g(x)\,dx$ diverges, then so does $\int_{a}^{\infty} f(x)\,dx$. This is called the **Comparison Test for Improper Integrals.**

Example 10: Is $\int_{2}^{\infty} \frac{1}{x^2+3x+2}\,dx$ convergent?

Solution: Because $\frac{1}{x^2} \geq \frac{1}{x^2+3x+2} \geq 0$ and $\int_{2}^{\infty} \frac{1}{x^2+3x+2}\,dx$ is convergent. (This is one of the integrals of the type $\int \frac{1}{x^p}\,dx$ with $p > 1$.)

✏️ **CRITICAL POINT**

As a practical matter, the p-Test is always a good tool to use when using the Comparison Test for Improper Integrals and part of the integrand to be tested contains a polynomial.

Example 11: Is $\int_{3}^{\infty} \frac{3+\cos(x)}{x-2}\,dx$ convergent?

Solution: Because $\frac{3+\cos(x)}{x-2} \geq \frac{1}{x} \geq 0$ and $\int_{3}^{\infty} \frac{1}{x}\,dx$ is divergent, therefore, $\int_{3}^{\infty} \frac{3+\cos(x)}{x-2}\,dx$ is divergent.

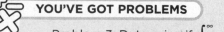 **YOU'VE GOT PROBLEMS**

Problem 3: Determine if $\int_{2}^{\infty} \frac{2}{\sqrt[3]{x^4+3x^2+1}}\,dx$ converges.

The Least You Need to Know

- When evaluating integrals with infinite bounds of integration, use a limit to determine the value of the integral, if it exists.

- Look for points of discontinuity on a finite interval, either at an endpoint of the interval or in the interior of the interval. If a discontinuity exists, use a limit to evaluate the integral.

- The *p*-Test is a good tool for comparing an unknown improper integral to a known improper integral.

Parametric Equations

We can treat some applications as one value that is a function of a second value (e.g., cost is a function of the number of items produced); we sketch and analyze the relationship as such. There are other applications that are more involved, even though they seem to be as straightforward as quantity versus cost relationship.

Some relationships cannot be written as simple statements such as y is a function of x. We like those because they are more straightforward to deal with than are other relationships. However, we'll be tackling these more complicated situations in this chapter. We begin with parametric equations.

If you need to, this would be a good time to go back to Chapter 1 and look over the material on parametric functions.

In This Chapter

- Working with first and second derivatives of parametric curves
- Computing the arc length of a parametric curve

First and Second Derivatives of Parametric Curves

Remember the Chain Rule for derivatives from Calculus I?

If …

- y is a function of u.

- u is a function of v.

- v is a function of x.

… then the derivative of y with respect to x is $\frac{dy}{dx} = \frac{dy}{du} \times \frac{du}{dv} \times \frac{dv}{dx}$. If nothing else makes sense about this statement, you should appreciate how the multiplication of fractions yields the desired derivative, $\frac{dy}{dx}$. (Think of how you cancel common factors when you multiply fractions.)

With parametric equations, the definitions of the functions are slightly different. Both the independent variable (usually x) and the dependent variable (usually y) are now both dependent on a third variable (for the moment, we'll call t). This is written out as: $x = f(t)$ and $y = g(t)$. What if we want to determine the value of $\frac{dy}{dx}$? We know that $f'(t) = \frac{dx}{dt}$ and that $g'(t) = \frac{dy}{dt}$. If we look at the fractional statement written, it makes sense that $\frac{dy}{dx} = \frac{\frac{dy}{dt}}{\frac{dx}{dt}}$.

> ✏️ **CRITICAL POINT**
>
> If x and y are both functions of an independent variable t, then $\frac{dy}{dx} = \frac{\frac{dy}{dt}}{\frac{dx}{dt}}$.

Example 1: Find the value of $\frac{dy}{dx}$ if $x = t^2 + 3t + 1$, and $y = 4t - 7$.

Solution: $\frac{dx}{dt} = 2t + 3$ and $\frac{dy}{dt} = 4$. Therefore, $\frac{dy}{dx} = \frac{4}{2t+3}$.

Example 2: Find the equation of the line tangent to the graph that has an equation: $x = 4\cos(t)$ and $y = 5\sin(t)$ when $t = \frac{\pi}{6}$.

Solution:

1. To write the equation of the line tangent to this ellipse, we'll need to know the point of tangency as well as the slope of the tangent line.

2. The point of tangency is $x = 4\cos\left(\frac{\pi}{6}\right) = 2\sqrt{3}$ and $y = 5\sin\left(\frac{\pi}{6}\right) = \frac{5}{2}$.

3. The slope of the tangent line at $t = \frac{\pi}{6}$ can be found using the derivative for parametric equations. $\frac{dx}{dt} = -4\sin(t)$ and $\frac{dy}{dt} = 5\cos(t)$. At $t = \frac{\pi}{6}$, $\frac{dx}{dt} = -2$ and $\frac{dy}{dt} = \frac{5\sqrt{3}}{2}$ so $\frac{dy}{dx} = \frac{\frac{5\sqrt{3}}{2}}{-2} = \frac{-5\sqrt{3}}{4}$.

4. To make life easy for ourselves, use the point-slope form for the line of an equation to get $y - \frac{5}{2} = \frac{-5\sqrt{3}}{4}\left(x - 2\sqrt{3}\right)$ as the tangent line.

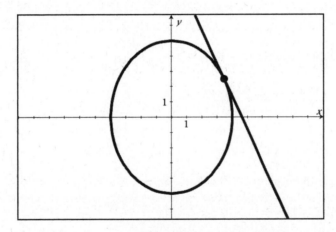

Figure 10.1
The graph of the ellipse $x = 4\cos(t)$ and $y = 5\sin(t)$ with tangent line drawn at the point
$$\left(2\sqrt{3}, \tfrac{5}{2}\right).$$

Example 3: The Lissajous curve $x = \sin(t)$ and $y = \sin(2t)$ $(0 \le t \le 2\pi)$ crosses itself at the origin. Find the equations of the lines tangent to this curve at the origin.

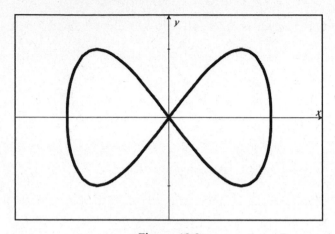

Figure 10.2

The graph of the Lissajous curve with equation $x = \sin(t)$ and $y = \sin(2t)$.

Solution:

1. We know the point of tangency is (0,0). All we need to do is find the derivative and the slopes of the tangent lines. $\frac{dx}{dt} = \cos(t)$ and $\frac{dy}{dt} = 2\cos(2t)$ so $\frac{dy}{dx} = \frac{2\cos(2t)}{\cos(t)}$.

2. To determine the slopes of the tangent lines, we need to determine when the graph crosses through the origin. We'll do this by working with the equation for x.

3. We know that $\sin(t) = 0$ when $t = 0$, π, and 2π.

4. At $t = 0$, $y = 0$; at $t = \pi$, $y = 0$; and at $t = 2\pi$, $y = 0$.

5. The slope of the tangent line at $t = 0$ is $\frac{dy}{dx} = \frac{2\cos(0)}{\cos(0)} = 2$.

6. The equation of one tangent line is $y = 2x$.

7. The slope of the tangent line at $t = \pi$ is $\frac{dy}{dx} = \frac{2\cos(2\pi)}{\cos(\pi)} = -2$.

8. The equation of a second tangent line is $y = -2x$.

9. The slope of the tangent line at $t = 2\pi$ is $\frac{dy}{dx} = \frac{2\cos(4\pi)}{\cos(2\pi)} = 2$.

10. The tangent line is again $y = 2x$.

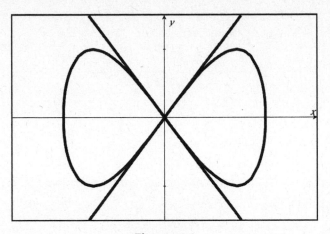

Figure 10.3
The graph of the Lissajous curve with equation $x = \sin(t)$ and $y = \sin(2t)$ with tangent lines $y = 2x$ and $y = -2x$ drawn.

Example 4: Find the slope of the curve denoted by $x = e^{-2t} \cos(t)$ and $y = e^{-t}\sin(2t)$ at $t = 0$.

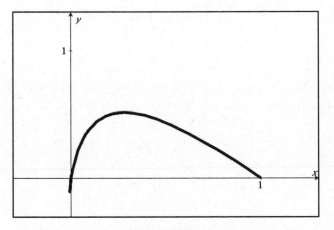

Figure 10.4
The graph of the curve denoted by $x = e^{-2t}\cos(t)$ and $y = e^{-t}\sin(2t)$.

Solution:

1. We need to compute $\frac{dx}{dt}$ and $\frac{dy}{dt}$. $\frac{dx}{dt} = -2e^{-2t}\cos(t) - e^{-2t}\sin(t)$ and $\frac{dy}{dt} = -e^{-t}\sin(2t) + 2e^{-t}\cos(2t)$.

2. This results in $\frac{dy}{dx} = \frac{-e^{-t}\sin(2t) + 2e^{-t}\cos(2t)}{-2e^{-2t}\cos(t) - e^{-2t}\sin(t)}$.

3. When $t = 0$, $\frac{dy}{dx} = \frac{-(1)(0) + 2(1)(1)}{-2(1)(1) - (1)(0)} = \frac{2}{-2} = -1$.

Example 5: Determine when the tangent lines for the curve denoted by the functions $x = 2t^2 - 1$ and $y = t^3 - 2t$ are horizontal and vertical.

Solution:

1. The slope of a horizontal line is 0. This will happen when $\frac{dy}{dt} = 0$ and $\frac{dx}{dt}$ is not 0.

2. $\frac{dy}{dt} = 3t^2 - 2$ and $\frac{dx}{dt} = 4t$. $3t^2 - 2 = 0$ when $t = \pm\frac{\sqrt{6}}{3}$.

3. When $t = \pm\frac{\sqrt{6}}{3}$, $\frac{dx}{dt} = \pm\frac{4\sqrt{6}}{3}$.

4. The slope of a vertical is undefined and this will occur when $\frac{dx}{dt} = 0$.

5. This happens when $t = 0$.

Example 6: Determine the coordinates of the point when the function $x = 50t$ and $y = 40t - 10t^2$ reaches its maximum value.

Solution:

1. The maximum value for the function occurs when $\frac{dy}{dx} = 0$.

2. $\frac{dx}{dt} = 50$ and $\frac{dy}{dt} = 40 - 20t$.

3. $\frac{dy}{dx} = \frac{40 - 20t}{50}$, and this equals 0 when $t = 2$.

4. $x(2) = 100$ and $y(2) = 80 - 40 = 40$.

5. The maximum value of the function is 40.

YOU'VE GOT PROBLEMS

Problem 1: Determine the equation of the line tangent to $x = e^t\tan(2t)$ and $y = e^{2t}\sec(t)$ at $t = 0$.

Now that we've practiced getting the first derivative of a parametrically defined function, let's take a look at the second derivative. $f''(x) = \frac{d\left(\frac{dy}{dx}\right)}{dx}$ in the traditional derivative notation. In parametric notation, $\frac{dy}{dx} = \frac{\frac{dy}{dt}}{\frac{dx}{dt}}$, so $\frac{d\left(\frac{dy}{dx}\right)}{dx} = \frac{\frac{d\left(\frac{dy}{dx}\right)}{dt}}{\frac{dx}{dt}}$.

BE AWARE

The second derivative in parametric form is *not* $\frac{\frac{d^2y}{dt^2}}{\frac{d^2x}{dt^2}}$.

Example 7: Determine the concavity of the Lissajous curve (Example 3) at $t = \frac{\pi}{6}$.

Solution:

1. We found $\frac{dy}{dx} = \frac{2\cos(2t)}{\cos(t)}$ in Example 3.

2. The second derivative is $\frac{\frac{d\left(\frac{dy}{dx}\right)}{dt}}{\frac{dx}{dt}} = \frac{\frac{\cot(t)(-4\sin(2t))-(2\cos(2t))(-\sin(t))}{\cos^2(t)}}{\cos(t)} = \frac{-4\cos(t)\sin(2t)+2\cos(2t)\sin(t)}{\cos^3(t)}$.

3. At $t = \frac{\pi}{6}$, $\frac{d^2y}{dx^2} = \frac{-4\cos\left(\frac{\pi}{6}\right)\sin\left(\frac{\pi}{3}\right)+2\cos\left(\frac{\pi}{3}\right)\sin\left(\frac{\pi}{6}\right)}{\cos^3\left(\frac{\pi}{6}\right)} = \frac{-4\left(\frac{\sqrt{3}}{2}\right)\left(\frac{\sqrt{3}}{2}\right)+2\left(\frac{1}{2}\right)\left(\frac{1}{2}\right)}{\left(\frac{\sqrt{3}}{2}\right)^3} = \frac{-3+\frac{1}{2}}{\frac{3\sqrt{3}}{8}} = \frac{-20}{3\sqrt{3}}$.

4. Because the second derivative is negative, the curve is concave down.

Example 8: Determine the concavity of the curve denoted by $x = 2t^2 - 1$ and $y = t^3 - 2t$ at $t = 3$.

Solution:

1. $\frac{dy}{dx} = \frac{3t^2 - 2}{4t}$

2. The second derivative is $\frac{\frac{d\left(\frac{dy}{dx}\right)}{dt}}{\frac{dx}{dt}} = \frac{\frac{4t(6t)-4(3t^2-2)}{16t^2}}{4t} = \frac{24t^2-12t^2+8}{64t^3} = \frac{12t^2+8}{64t^3} = \frac{3t^2+2}{16t^3}$.

3. At $t = 3$, $\frac{d^2y}{dx^2} = \frac{3(3)^2+2}{16(3)^3} = \frac{29}{432}$.

4. Because the second derivative is positive, the curve is concave up.

YOU'VE GOT PROBLEMS

Problem 2: Determine the value of the second derivative of $x = 4\cos(3t)$ and $y = 3\sin(4t)$ at $t = \frac{\pi}{2}$.

Arc Length of a Parametric Curve

We saw in Chapter 5 that the length of an arc is $\int_a^b \sqrt{(dx)^2 + (dy)^2}$. Also in Chapter 5, we simplified the radicand by factoring out the term $(dx)^2$. For parametric curves, we'll multiply and divide the radicand by $(dx)^2$ making the arc length formula $\int_a^b \sqrt{\left(\frac{dx}{dt}\right)^2 + \left(\frac{dy}{dt}\right)^2}\, dt$.

Example 9: Find the circumference of the circle defined by $x = a\cos(t)$ and $y = a\sin(t)$.

Solution: The bounds of integration are 0 and 2π. $\frac{dx}{dt} = -a\sin(t)$ and $\frac{dy}{dt} = a\cos(t)$. The length of the arc is $\int_0^{2\pi} \sqrt{(-a\sin(t))^2 + (a\cos(t))^2}\, dt = \int_0^{2\pi} \sqrt{a^2\sin^2(t) + a^2\cos^2(t)}\, dt =$
$\int_0^{2\pi} \sqrt{a^2\left(\sin^2(t) + \cos^2(t)\right)}\, dt = \int_0^{2\pi} \sqrt{a^2}\, dt = \int_0^{2\pi} a\, dt = \left. at \right|_0^{2\pi} = 2\pi a - 0 = 2\pi a.$

Example 10: Find the length of the arc for the curve defined by $x = e^t\cos(t)$ and $y = e^t\sin(t)$ on $[0,5]$.

Solution: $\frac{dx}{dt} = e^t\cos(t) - e^t\sin(t) = e^t\left(\cos(t) - \sin(t)\right)$ and

$\frac{dy}{dt} = e^t\sin(t) + e^t\cos(t) = e^t\left(\sin(t) + \cos(t)\right) \cdot \left(\frac{dx}{dt}\right)^2 = e^{2t}\left(\cos(t) - \sin(t)\right)^2 =$

$e^{2t}\left(\cos^2(t) - 2\sin(t)\cos(t) + \sin^2(t)\right) = e^{2t}\left(1 - \sin(2t)\right).$

$\left(\frac{dy}{dt}\right)^2 = e^{2t}\left(\sin(t) + \cos(t)\right)^2 = e^{2t}\left(\sin^2(t) + 2\sin(t)\cos(t) + \cos^2(t)\right) = e^{2t}\left(1 + \sin(2t)\right)$

Therefore, the length of the arc is $\int_0^5 \sqrt{e^{2t}(1 - \sin 2(t)) + e^{2t}(1 + \sin(2t))}\, dt =$

$\int_0^5 \sqrt{e^{2t}\left(1 - \sin 2(t) + 1 + \sin(2t)\right)}\, dt = \int_0^5 \sqrt{2e^{2t}}\, dt = \sqrt{2}\int_0^5 e^t\, dt = \left. \sqrt{2}\, e^t \right|_0^5 = \sqrt{2}\left(e^5 - 1\right).$

Example 11: Determine the distance around the circumference of the ellipse with equation $x = 4\cos(t)$ and $y = 5\sin(t)$.

Solution: The bounds of integration will be 0 to 2π and the length of the arc is $\int_0^{2\pi} \sqrt{16\sin^2(t) + 25\cos^2(t)}\, dt$. This is an example of an integral that we do not know how to evaluate, so we'll use the calculator to get the answer 28.3617.

YOU'VE GOT PROBLEMS

Problem 3: Find the length of the arc formed by $x = \tan^{-1}(t)$ and $y = \ln\left(\sqrt{1+t^2}\right)$ on [0, 1].

The Least You Need to Know

- If $x = f(t)$ and $y = g(t)$, then $\frac{dy}{dx} = \frac{g'(t)}{f'(t)}$.

- The second derivative $\frac{d\left(\frac{dy}{dx}\right)}{dx} = \frac{\frac{d\left(\frac{dy}{dx}\right)}{dt}}{\frac{dx}{dt}}$.

- The length of an arc of a curve defined parametrically is $\int_a^b \sqrt{\left(\frac{dx}{dt}\right)^2 + \left(\frac{dy}{dt}\right)^2}\, dt$.

Polar Coordinates

As we talked about in Chapter 1, the study of mathematics would be quite different if Descartes had used a series of concentric circles when he tried to determine the location of the bug on the ceiling rather than the series of parallel and perpendicular lines that became the Rectangular Coordinate System (or better still, the Cartesian Coordinate System). We will take a little time in this chapter to examine some of the basics of calculus—slopes of tangent lines, concavity, length of an arc of a polar curve, and the area under a curve.

Before reading on, you might want to go back and review the material on polar coordinates in Chapter 1.

In This Chapter

- Computing the slope of the line tangent to a polar curve
- Determining the length of an arc of a polar curve
- Calculating area under a polar curve

Slope of the Tangent Line

As we just saw in Chapter 10 and the study of parametrically defined functions, we still think of the slope of the tangent in terms of the first derivative, $\frac{dy}{dx}$. We need to recall how the polar and Rectangular Coordinate systems are related to one another. The conversion from polar to rectangular is $x = r\cos(\theta)$ and $y = r\sin(\theta)$. Consequently, $\frac{dy}{dx} = \frac{\frac{d(r\sin(\theta))}{d\theta}}{\frac{d(r\cos(\theta))}{d\theta}}$.

Example 1: Find the slope of the line tangent to $r = 3\cos(2\theta)$ at $\theta = \frac{\pi}{6}$.

Solution:

1. The slope of the tangent line is $\frac{dy}{dx} = \frac{\frac{d(r\sin(\theta))}{d\theta}}{\frac{d(r\cos(\theta))}{d\theta}}$.

2. Use the product rule to compute both the numerator and denominator for $\frac{dy}{dx}$.

3. The numerator of the derivative, $\frac{d(r\sin(\theta))}{d\theta} = \frac{d((3\cos(2\theta))\sin(\theta))}{d\theta} = -6\sin(2\theta)\sin(\theta) + 3\cos(2\theta)\cos(\theta)$ while the denominator of the derivative is $\frac{d(r\cos(\theta))}{d\theta} = \frac{d((3\cos(2\theta))\cos(\theta))}{d\theta} = -6\sin(2\theta)\cos(\theta) - 3\cos(2\theta)\sin(\theta)$.

4. We now have $\frac{dy}{dx} = \frac{-6\sin(2\theta)\sin(\theta) + 3\cos(2\theta)\cos(\theta)}{-6\sin(2\theta)\cos(\theta) - 3\cos(2\theta)\sin(\theta)}$.

5. Evaluate this expression at $\theta = \frac{\pi}{6}$, $\frac{dy}{dx} = \frac{-6\sin\left(\frac{\pi}{3}\right)\sin\left(\frac{\pi}{6}\right) + 3\cos\left(\frac{\pi}{3}\right)\cos\left(\frac{\pi}{6}\right)}{-6\sin\left(\frac{\pi}{3}\right)\cos\left(\frac{\pi}{6}\right) - 3\cos\left(\frac{\pi}{3}\right)\sin\left(\frac{\pi}{6}\right)} = \frac{\left(-3\sqrt{3}\right)\left(\frac{1}{2}\right) + \left(\frac{3}{2}\right)\left(\frac{\sqrt{3}}{2}\right)}{\left(-3\sqrt{3}\right)\left(\frac{\sqrt{3}}{2}\right) - \left(\frac{3}{2}\right)\left(\frac{1}{2}\right)} = \frac{-6\sqrt{3} + 3\sqrt{3}}{-18 - 3} = \frac{-3\sqrt{3}}{-21} = \frac{\sqrt{3}}{7}$.

Example 2: Write the equation of the line tangent to $r = 3\cos(2\theta)$ at $\theta = \frac{\pi}{6}$.

Solution:

1. We know the slope of the line is $= \frac{\sqrt{3}}{7}$.

2. We need to determine the coordinates of the point through which the point passes. When $\theta = \frac{\pi}{6}$, $r = 3\cos\left(\frac{\pi}{3}\right) = \frac{3}{2}$.

3. Convert polar to rectangular coordinates: $x = \frac{3}{2}\cos\left(\frac{\pi}{6}\right) = \left(\frac{3}{2}\right)\left(\frac{\sqrt{3}}{2}\right) = \frac{3\sqrt{3}}{4}$ and $y = \frac{3}{2}\sin\left(\frac{\pi}{6}\right) = \left(\frac{3}{2}\right)\left(\frac{1}{2}\right) = \frac{3}{4}$.

4. Consequently, the equation of the line tangent to $r = 3\cos(2\theta)$ at $\theta = \frac{\pi}{6}$ is $y - \frac{3}{4} = \frac{\sqrt{3}}{7}\left(x - \frac{3\sqrt{3}}{4}\right)$.

Example 3: Write the equation of the lines tangent to $r = 3\cos(2\theta)$ when the graph passes through the pole (the point when the radius is equal to zero).

Solution:

1. There is a little bit of algebra to do here. When solving the equation $3\cos(2\theta) = 0$, we are working with the domain $0 \le \theta \le 2\pi$. However, our equation contains $\cos(2\theta)$, so we need to consider the values for 2θ.

2. Double the values for the interval $0 \le \theta \le 2\pi$ to get $0 \le 2\theta \le 4\pi$.

3. On this interval, $\cos(2\theta)$ when $2\theta = \frac{\pi}{2}, \frac{3\pi}{2}, \frac{5\pi}{2}, \frac{7\pi}{2}$, so $\theta = \frac{\pi}{4}, \frac{3\pi}{4}, \frac{5\pi}{4}, \frac{7\pi}{4}$.

4. The point of tangency in each case will be the origin $(0,0)$.

5. The slope of the tangent lines at each of these values is:

   - At $\theta = \frac{\pi}{4}$, $\frac{dy}{dx} = \frac{-6\sin\left(\frac{\pi}{2}\right)\sin\left(\frac{\pi}{4}\right) + 3\cos\left(\frac{\pi}{2}\right)\cos\left(\frac{\pi}{4}\right)}{-6\sin\left(\frac{\pi}{2}\right)\cos\left(\frac{\pi}{4}\right) - 3\cos\left(\frac{\pi}{2}\right)\sin\left(\frac{\pi}{4}\right)} = \frac{-6(1)\left(\frac{\sqrt{2}}{2}\right) + 3(0)\left(\frac{\sqrt{2}}{2}\right)}{-6(1)\left(\frac{\sqrt{2}}{2}\right) - 3(0)\left(\frac{\sqrt{2}}{2}\right)} = \frac{-3\sqrt{2}}{-3\sqrt{2}} = 1$. The tangent line has equation $y = x$.

   - At $\theta = \frac{3\pi}{4}$, $\frac{dy}{dx} = \frac{-6\sin\left(\frac{3\pi}{2}\right)\sin\left(\frac{3\pi}{4}\right) + 3\cos\left(\frac{3\pi}{2}\right)\cos\left(\frac{3\pi}{4}\right)}{-6\sin\left(\frac{3\pi}{2}\right)\cos\left(\frac{3\pi}{4}\right) - 3\cos\left(\frac{3\pi}{2}\right)\sin\left(\frac{3\pi}{4}\right)} = \frac{-6(-1)\left(\frac{\sqrt{2}}{2}\right) + 3(0)\left(\frac{-\sqrt{2}}{2}\right)}{-6(-1)\left(\frac{-\sqrt{2}}{2}\right) - 3(0)\left(\frac{\sqrt{2}}{2}\right)} = \frac{3\sqrt{2}}{-3\sqrt{2}} = -1$. The tangent line has equation $y = -x$.

   - At $\theta = \frac{5\pi}{4}$, $\frac{dy}{dx} = \frac{-6\sin\left(\frac{5\pi}{2}\right)\sin\left(\frac{5\pi}{4}\right) + 3\cos\left(\frac{5\pi}{2}\right)\cos\left(\frac{5\pi}{4}\right)}{-6\sin\left(\frac{5\pi}{2}\right)\cos\left(\frac{5\pi}{4}\right) - 3\cos\left(\frac{5\pi}{2}\right)\sin\left(\frac{5\pi}{4}\right)} = \frac{-6(1)\left(\frac{-\sqrt{2}}{2}\right) + 3(0)\left(\frac{-\sqrt{2}}{2}\right)}{-6(1)\left(\frac{-\sqrt{2}}{2}\right) - 3(0)\left(\frac{-\sqrt{2}}{2}\right)} = \frac{3\sqrt{2}}{3\sqrt{2}} = 1$. The tangent line has equation $y = x$.

   - At $\theta = \frac{7\pi}{4}$, $\frac{dy}{dx} = \frac{-6\sin\left(\frac{7\pi}{2}\right)\sin\left(\frac{7\pi}{4}\right) + 3\cos\left(\frac{7\pi}{2}\right)\cos\left(\frac{7\pi}{4}\right)}{-6\sin\left(\frac{7\pi}{2}\right)\cos\left(\frac{7\pi}{4}\right) - 3\cos\left(\frac{7\pi}{2}\right)\sin\left(\frac{7\pi}{4}\right)} = \frac{-6(-1)\left(\frac{-\sqrt{2}}{2}\right) + 3(0)\left(\frac{\sqrt{2}}{2}\right)}{-6(-1)\left(\frac{\sqrt{2}}{2}\right) - 3(0)\left(\frac{-\sqrt{2}}{2}\right)} = \frac{-3\sqrt{2}}{3\sqrt{2}} = -1$. The tangent line has equation $y = -x$.

Example 4: When is the tangent line to the graph $r = 3\cos(2\theta)$ vertical?

Solution:

1. The slope of a vertical line is undefined, so we need to find those values of θ for which $\frac{dx}{d\theta} = 0$.

2. Solve $-6\sin(2\theta)\cos(\theta) - 3\cos(2\theta)\sin(\theta) = 0$ by first using the double angle identities for sine and cosine and then removing the common factor, $-3\sin(\theta)(6\cos^2(\theta) - 1) = 0$.

3. Set each of the factors equal to 0 and solve. $-3\sin(\theta) = 0$ when $\theta = 0, \pi, 2\pi$ and $\cos(\theta) + \sin(\theta) = 0$ when $\theta = 1.1503, 1.9913, 4.2919, 5.1329$.

✏️ **CRITICAL POINT**

When solving trigonometric equations with the input values of the form $n\theta$ and the domain for θ is $0 \leq \theta \leq 2\pi$, you must solve the equation over the interval $0 \leq n\theta \leq 2n\pi$.

Equations of the form $r = a\cos(n\theta)$ and $r = a\sin(n\theta)$ are called roses. The number of petals on the rose is ...

- n if n is an odd integer.

- $2n$ if n is an even integer.

Two other popular polar curves are as follows:

- *Cardioid* $(r = a \pm a\cos(\theta)$ and $r = a \pm a\sin(\theta))$

- *Limaçon* $(r = a \pm b\cos(\theta)$ and $r = a \pm b\sin(\theta))$

Example 5: Find the slope of the line tangent to $r = 4 + 2\sin(\theta)$ at $\theta = \frac{\pi}{3}$.

Solution:

1. As we saw in Example 1, we'll need to use the product rule to compute $\frac{dy}{d\theta}$ and $\frac{dx}{d\theta}$.

2. Convert to rectangular form, $y = (4 + 2\sin(\theta))\sin(\theta) = 4\sin(\theta) + 2\sin^2(\theta)$, so that
$$\frac{dy}{d\theta} = 4\cos(\theta) + 4\sin(\theta)\cos(\theta) = 4\cos(\theta) + 2\sin(2\theta).$$

3. In rectangular form, $x = (4 + 2\sin(\theta))\cos(\theta) = 4\cos(\theta) + 2\sin(\theta)\cos(\theta) = 4\cos(\theta) + \sin(2\theta)$,
so that $\frac{dx}{d\theta} = -4\sin(\theta) + 2\cos(2\theta)$.

4. The slope of the tangent line is $\frac{dy}{dx} = \frac{4\cos(\theta) + 2\sin(2\theta)}{-4\sin(\theta) + 2\cos(2\theta)}$.

5. At $\theta = \frac{\pi}{3}$, the slope of the tangent line is $\frac{dy}{dx} = \frac{4\cos\left(\frac{\pi}{3}\right) + 2\sin\left(\frac{2\pi}{3}\right)}{-4\sin\left(\frac{\pi}{3}\right) + 2\cos\left(\frac{2\pi}{3}\right)} = \frac{2 + \sqrt{3}}{-2\sqrt{3} - 1}$.

Example 6: Determine the equations of the lines tangent to $r = 2 + 4\sin(\theta)$ when the graph passes through the pole.

Solution:

1. The graph passes through the pole (rectangular coordinates (0,0)) when $2 + 4\sin(\theta) = 0$.

2. Solve this equation, $\sin(\theta) = \frac{-1}{2}$ so that $\theta = \frac{7\pi}{6}, \frac{11\pi}{6}$.

3. The derivative $\dfrac{dy}{dx} = \dfrac{\frac{d((2+4\sin(\theta))\sin(\theta))}{d\theta}}{\frac{d((2+4\sin(\theta))\cos(\theta))}{d\theta}} = \dfrac{\frac{d(2\sin(\theta)+4\sin^2(\theta))}{d\theta}}{\frac{d(2\cos(\theta)+4\sin(\theta)\cos(\theta))}{d\theta}} = \dfrac{\frac{d(2\sin(\theta)+4\sin^2(\theta))}{d\theta}}{\frac{d(2\cos(\theta)+2\sin(2\theta))}{d\theta}} =$

    $\dfrac{2\cos(\theta)+8\sin(\theta)\cos(\theta)}{-2\sin(\theta)+4\cos(2\theta)} = \dfrac{2\cos(\theta)(1+4\sin(\theta))}{2(-\sin(\theta)+2\cos(2\theta))} = \dfrac{\cos(\theta)(1+4\sin(\theta))}{-\sin(\theta)+2\cos(2\theta)}$.

4. When $\theta = \frac{7\pi}{6}$, $\dfrac{dy}{dx} = \dfrac{\cos(\frac{7\pi}{6})(1+4\sin(\frac{7\pi}{6}))}{-\sin(\frac{7\pi}{6})+2\cos(\frac{7\pi}{3})} = \dfrac{(\frac{-\sqrt{3}}{2})(1+4(\frac{-1}{2}))}{-(\frac{-1}{2})+2(\frac{1}{2})} = \dfrac{(\frac{-\sqrt{3}}{2})(1-2)}{\frac{1}{2}+1} = \dfrac{\frac{\sqrt{3}}{2}}{\frac{3}{2}} = \dfrac{\sqrt{3}}{3}$.

5. The equation of the tangent line is $y = \frac{\sqrt{3}}{3}x$.

6. When $\theta = \frac{11\pi}{6}$, $\dfrac{dy}{dx} = \dfrac{\cos(\frac{11\pi}{6})(1+4\sin(\frac{11\pi}{6}))}{-\sin(\frac{11\pi}{6})+2\cos(\frac{11\pi}{3})} = \dfrac{(\frac{\sqrt{3}}{2})(1+4(\frac{-1}{2}))}{-(\frac{-1}{2})+2(\frac{1}{2})} = \dfrac{(\frac{\sqrt{3}}{2})(1+-2)}{\frac{1}{2}+1} = \dfrac{\frac{-\sqrt{3}}{2}}{\frac{3}{2}} = \dfrac{-\sqrt{3}}{3}$.

7. The equation of the tangent line is $y = \frac{-\sqrt{3}}{3}x$.

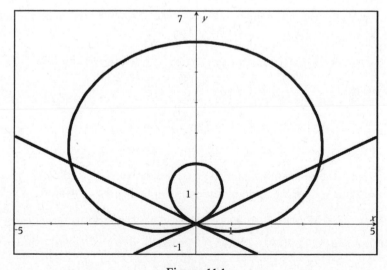

Figure 11.1
The graph of the limaçon $r = 2 + 4\sin(\theta)$ with the tangent lines drawn at the pole.

YOU'VE GOT PROBLEMS

Problem 1: Find the equation of the line tangent to $r = 3 - 5\cos(\theta)$ at the point when $\theta = \frac{2\pi}{3}$.

Computing the second derivative in polar coordinates is as challenging as it was with parametrically defined functions: $\frac{d^2y}{dx^2} = \frac{d\left(\frac{dy}{dx}\right)}{dx} = \frac{\frac{d\left(\frac{dy}{dx}\right)}{d\theta}}{\frac{dx}{d\theta}}$. Just know up front that this is going to get sloppy and that you won't need to do this often.

Example 7: Find the second derivative of the polar curve $r = 2 + 4\sin(\theta)$.

Solution:

1. We know from Example 6 that $\frac{dy}{dx} = \frac{\cos(\theta)\left(1 + 4\sin(\theta)\right)}{-\sin(\theta) + 2\cos(2\theta)}$.

$$\frac{d^2y}{dx^2} = \frac{\frac{d\left(\frac{dy}{dx}\right)}{dt}}{\frac{dx}{dt}} = \frac{\frac{d\left(\frac{\cos(\theta)\left(1+4\sin(\theta)\right)}{-\sin(\theta)+2\cos(2\theta)}\right)}{dt}}{\frac{dx}{dt}} = \frac{\frac{\left(-\sin(\theta)+2\cos(2\theta)\right)\left[-\sin(\theta)(1+4\sin(\theta))+\cos(\theta)(4\cos(\theta))\right]-\left(-\cos(\theta)-4\sin(2\theta)\right)\left[\cos(\theta)(1+4\sin(\theta))\right]}{\left(-\sin(\theta)+2\cos(2\theta)\right)^2}}{-2\sin(\theta)+4\cos(2\theta)} =$$

$$\frac{\left(-\sin(\theta)+2\cos(2\theta)\right)\left[-\sin(\theta)(1+4\sin(\theta))+\cos(\theta)(4\cos(\theta))\right]-\left(-\cos(\theta)-4\sin(2\theta)\right)\left[\cos(\theta)(1+4\sin(\theta))\right]}{2\left(-\sin(\theta)+2\cos(2\theta)\right)^3}$$

2. It might be possible to simplify this, but I don't really care to mess with it. Now, let's take a look at the graphing calculator. I define the original function as $r(t)$ because I am too lazy to write r(theta).

3. I'll use the derivative command to get the first derivative and then find the second derivative using the formula $\frac{d^2y}{dx^2} = \frac{\frac{d\left(\frac{\cos(\theta)\left(1+4\sin(\theta)\right)}{-\sin(\theta)+2\cos(2\theta)}\right)}{dt}}{\frac{dx}{dt}}$. One of the "joys" of using the graphing calculator to get the symbolic answer for the derivative is that you are never quite sure which trigonometric identity it will choose to use.

4. Notice that the denominator for the first derivative is $2\cos^2(t) - \sin(t)(2\sin(t) + 1)$ and the denominator we got when doing the problem by hand was $-\sin(\theta) + 2\cos(2\theta)$.

5. Let's take the calculator's response, manipulate it, and show that it is the same as ours.

$2\cos^2(t) - \sin(t)(2\sin(t) + 1) = 2\cos^2(t) - 2\sin^2(t) - \sin(t) = 2(\cos^2(t) - \sin^2(t)) - \sin(t)$

6. One of the forms for the identity cos(2*t*) is cos²(*t*) – sin²(*t*), so 2(cos²(*t*) – sin² (*t*)) – sin(*t*) = 2cos(2*t*) – sin(*t*), which is what we got. If you would like to monkey with the numerator to show that they are the same, please do so.

$$r(t) := 2 + 4 \cdot \sin(t) \hspace{6cm} Done$$

$$\frac{d}{dt}\big(r(t) \cdot \sin(t)\big)$$
$$\frac{d}{dt}\big(r(t) \cdot \cos(t)\big)$$

$$\frac{\big(4 \cdot \sin(t) + 1\big) \cdot \cos(t)}{2 \cdot \big(\cos(t)\big)^2 - \sin(t) \cdot \big(2 \cdot \sin(t) + 1\big)}$$

$$\frac{\dfrac{d}{dt}\left(\dfrac{\big(4 \cdot \sin(t) + 1\big) \cdot \cos(t)}{2 \cdot \big(\cos(t)\big)^2 - \sin(t) \cdot \big(2 \cdot \sin(t) + 1\big)} \right)}{\dfrac{d}{dt}\big(r(t) \cdot \cos(t)\big)}$$

$$\frac{8 \cdot \big(\cos(t)\big)^2 + \big(2 \cdot \sin(t) + 1\big) \cdot \big(4 \cdot \sin(t) + 1\big)}{2 \cdot \big(2 \cdot \big(\cos(t)\big)^2 - \sin(t) \cdot \big(2 \cdot \sin(t) + 1\big)\big)^3}$$

Figure 11.2

Define the original function and then compute the first and second derivatives on the TI-Nspire.

Length of an Arc of a Polar Curve

We saw in an earlier chapter that the length of an arc along the graph of a function f(*x*) is

$$L = \int_a^b \sqrt{1 + \left(\tfrac{dy}{dx}\right)^2}\, dx = \int_a^b \sqrt{\big(dx\big)^2 + \big(dy\big)^2}\,.$$ When the function was written in parametric form, the

length of the arc was computed as $L = \int_a^b \sqrt{\left(\tfrac{dx}{dt}\right)^2 + \left(\tfrac{dy}{dt}\right)^2}\, dt$. Now we'll take a look at the process for finding the length of an arc in polar form.

1. We take the integral $L = \int_a^b \sqrt{\big(dx\big)^2 + \big(dy\big)^2}$ and multiply and divide the integrand by the

   differential *d*θ to get $L = \int_a^b \sqrt{\left(\tfrac{dx}{d\theta}\right)^2 + \left(\tfrac{dy}{d\theta}\right)^2}\, d\theta$.

   $x = r\cos(\theta)$, $\tfrac{dx}{d\theta} = \tfrac{dr}{d\theta}\cos(\theta) - r\sin(\theta)$

2. This causes $\left(\tfrac{dx}{d\theta}\right)^2$ to equal $\left(\tfrac{dr}{d\theta}\cos(\theta) - r\sin(\theta)\right)^2 =$

   $\left(\tfrac{dr}{d\theta}\right)^2 \cos^2(\theta) - 2r\tfrac{dr}{d\theta}\cos(\theta)\sin(\theta) + r^2\sin^2(\theta)\,.$

3. In a similar fashion, $\left(\frac{dy}{d\theta}\right)^2 = \left(\frac{dr}{d\theta}\right)^2 \sin^2(\theta) + 2r\left(\frac{dr}{d\theta}\right)\sin(\theta)\cos(\theta) + r^2 \cos^2(\theta)$.

4. That doesn't look like a lot of fun but as they say in the infomercials, "But wait, there's more!"

$$\sqrt{\left(\frac{dx}{d\theta}\right)^2 + \left(\frac{dy}{d\theta}\right)^2} =$$

$$\sqrt{\left(\frac{dr}{d\theta}\right)^2 \cos^2(\theta) - 2r\frac{dr}{d\theta}\cos(\theta)\sin(\theta) + r^2 \sin^2(\theta) + \left(\frac{dr}{d\theta}\right)^2 \sin^2(\theta) + 2r\frac{dr}{d\theta}\cos(\theta)\sin(\theta) + r^2 \cos^2(\theta)}$$

5. The terms $-2r\left(\frac{dr}{d\theta}\right)\sin(\theta)\cos(\theta)$ and $2r\left(\frac{dr}{d\theta}\right)\sin(\theta)\cos(\theta)$ add to equal zero.

6. Simplify the rest of the radicand, $\sqrt{\left(\frac{dr}{d\theta}\right)^2 \left(\cos^2(\theta) + \sin^2(\theta)\right) + r^2 \left(\sin^2(\theta) + \cos^2(\theta)\right)} =$

$$\sqrt{\left(\frac{dr}{d\theta}\right)^2 + r^2}\ .$$

7. Therefore, $L = \int_a^b \sqrt{\left(\frac{dr}{d\theta}\right)^2 + r^2}\ d\theta$.

CRITICAL POINT

The length of an arc formed by a polar graph $r(\theta)$ on the interval $a \le \theta \le b$ is $0 \le \theta \le \frac{\pi}{4}$.

Example 8: Find the length of the arc of the rose $r = 3\cos(2\theta)$ on the interval $0 \le \theta \le \frac{\pi}{4}$.

Solution: $\frac{dr}{d\theta} = -6\sin(2\theta)$ so $L = \int_0^{\pi/4} \sqrt{36\sin^2(2\theta) + 9\cos^2(2\theta)}\ d\theta$. Use the graphing calculator to determine that the length is 3.633.

Example 9: Find the length around the entire graph of the rose $r = 3\cos(2\theta)$.

Solution: The section of the rose covered in the interval $0 \le \theta \le \frac{\pi}{4}$ is the upper portion of the first petal.

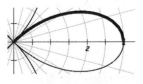

Figure 11.3
The upper section of the first petal of the rose $r = 3\cos(2\theta)$.

Therefore, the length of the distance around the entire rose is

$$L = 8\int_0^{\pi/4} \sqrt{36\sin^2(2\theta) + 9\cos^2(2\theta)}\, d\theta = 29.065.$$

Example 10: Find the length of the arc of the limaçon $r = 2 + 4\sin(\theta)$ on the interval $-\frac{\pi}{2} \le \theta \le \frac{\pi}{2}$.

Solution: $\frac{dr}{d\theta} = 4\cos(\theta)$, so $L = \int_{-\pi/2}^{\pi/2} \sqrt{16\cos^2(\theta) + (2 + 4\sin(\theta))^2}\, d\theta =$

$\int_{-\pi/2}^{\pi/2} \sqrt{16\cos^2(\theta) + 4 + 16\sin(\theta) + 16\sin^2(\theta)}\, d\theta = \int_{-\pi/2}^{\pi/2} \sqrt{20 + 16\sin(\theta)}\, d\theta$. Again, use the

calculator to compute this value. $L = \int_{-\pi/2}^{\pi/2} \sqrt{20 + 16\sin(\theta)}\, d\theta = 13.365.$

(As you can see from the sketch of the graph, this value represents half the way around the graph. The full perimeter is twice this number.)

CRITICAL POINT

There are times when you'll need to work with values of θ that are less than 0 in order to have a continuous interval on which to compute an integral.

YOU'VE GOT PROBLEMS

Problem 2: Determine the total distance around the graph of the rose $r = 4\sin(3\theta)$. (It will be worth your while to look at a picture of this graph on your calculator.)

Area Under a Curve

Recall that when we were first looking at area under a curve in the rectangular coordinate system, we partitioned the region and created rectangles whose area were easily computed. (We also looked at trapezoids and arcs of parabolas.) As we increased the number of rectangles in the interval, the width of each rectangle got smaller and smaller, leading us to conclude the product $f(x_i)dx_i$ represented the area of one for the ith rectangle.

We'll use a similar way to compute area in the polar coordinate system. Rather than having rectangles, though, we'll have sectors that will make up the partition. You might recall from your study of trigonometry that the area of a sector of a circle with radius r that contains θ radians is $\frac{1}{2}r^2\theta$. (Solve the proportion $\frac{A}{\pi r^2} = \frac{\theta}{2\pi}$ for A.) The argument we would use to go from the area of a rectangle to the Fundamental Theorem of Calculus also takes us from the area of a sector to the area under a curve in the polar coordinate system.

> **DEFINITION**
>
> The area under the polar graph $r(\theta)$ on the interval $a \le \theta \le b$ is $\int_a^b \frac{1}{2}r^2\,d\theta$.

Example 11: Find the area of the arc of the rose $r = 3\cos(2\theta)$ on the interval $0 \le \theta \le \frac{\pi}{4}$.

Solution: $A = \frac{1}{2}\int_0^{\pi/4} 9\cos^2(2\theta)d\theta = \frac{9}{2}\int_0^{\pi/4}\cos^2(2\theta)d\theta$.

Use the trigonometric identity $\cos(2\theta) = 2\cos^2(\theta) - 1$ to rewrite $\cos^2(\theta) = \frac{1}{2}\cos(2\theta) + \frac{1}{2}$.

$\frac{9}{2}\int_0^{\pi/4}\cos^2(2\theta)d\theta = \frac{9}{2}\int_0^{\pi/4}\frac{1}{2}\cos(4\theta) + \frac{1}{2}\,d\theta = \frac{9}{2}\left(\frac{1}{4}\sin(4\theta) - \theta\right)\Big|_0^{\pi/4} =$

$\frac{9}{4}\left[\left(\frac{1}{4}\sin(\pi) + \frac{\pi}{4}\right) - \left(\frac{1}{4}\sin(0) + 0\right)\right] = \frac{9\pi}{16}$. (Multiply this number 8 to get the complete area within the rose.)

Example 12: Find the area of the arc of the limaçon $r = 2 + 4\sin(\theta)$ on the interval $-\frac{\pi}{2} \le \theta \le \frac{\pi}{2}$.

Solution: $A = \frac{1}{2}\int_{-\pi/2}^{\pi/2}\left(2 + 4\sin(\theta)\right)^2 d\theta = \frac{1}{2}\int_{-\pi/2}^{\pi/2} 4 + 16\sin(\theta) + 16\sin^2(\theta)\,d\theta$.

Use the trigonometric identity $\cos(2\theta) = 1 - 2\sin^2(\theta)$ to rewrite $\sin^2(\theta) = \frac{1}{2} - \frac{1}{2}\cos(2\theta)$.

$$\frac{1}{2}\int_{-\pi/2}^{\pi/2} 4 + 16\sin(\theta) + 16\sin^2(\theta)\,d\theta = \frac{1}{2}\int_{-\pi/2}^{\pi/2} 4 + 16\sin(\theta) + 16\left(\frac{1}{2} - \frac{1}{2}\cos(2\theta)\right) d\theta =$$

$$\frac{1}{2}\int_{-\pi/2}^{\pi/2} 12 + 16\sin(\theta) - 8\cos(2\theta)\,d\theta = \frac{1}{2}\left(12\theta - 16\cos(\theta) - 4\sin(2\theta)\right)\Big|_{-\pi/2}^{\pi/2} =$$

$$\frac{1}{2}\left[\left(12\left(\frac{\pi}{2}\right) - 16\cos\left(\frac{\pi}{2}\right) - 4\sin(\pi)\right) - \left(12\left(\frac{-\pi}{2}\right) - 16\cos\left(\frac{-\pi}{2}\right) - 4\sin(-\pi)\right)\right] =$$

$$\frac{1}{2}\left[\left(6\pi - 0 - 0\right) - \left(-6\pi - 0 - 0\right)\right] = 6\pi$$

Example 13: Find the area of the inner loop of the limaçon $r = 2 + 4\sin(\theta)$.

Solution:

1. We need to determine a continuous interval that will form the inner loop (or half the inner loop—we can always double the answer). Solve $2 + 4\sin(\theta) = 0$ to get $\sin(\theta) = \frac{-1}{2}$ so that $\theta = \frac{7\pi}{6}, \frac{11\pi}{6}$.

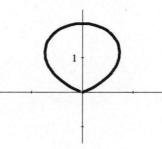

Figure 11.4
Inner loop of $r = 2 + 4\sin(\theta)$.

2. The area is $\frac{1}{2}\int_{7\pi/6}^{11\pi/6}\left(2 + 4\sin(\theta)\right)^2 d\theta$. The antiderivative will be the same as in Example 12, so $\frac{1}{2}\int_{7\pi/6}^{11\pi/6}\left(2 + 4\sin(\theta)\right)^2 d\theta = \frac{1}{2}\left(12\theta - 16\cos(\theta) - 4\sin(2\theta)\right)\Big|_{7\pi/6}^{11\pi/6} =$

$\frac{1}{2}\left[\left(12\left(\frac{11\pi}{6}\right) - 16\cos\left(\frac{11\pi}{6}\right) - 4\sin\left(\frac{11\pi}{3}\right)\right) - \left(12\left(\frac{7\pi}{6}\right) - 16\cos\left(\frac{7\pi}{6}\right) - 4\sin\left(\frac{7\pi}{3}\right)\right)\right]$.

3. This equals $\frac{1}{2}\left[\left(22\pi - 8\sqrt{3} + 2\sqrt{3}\right) - \left(14\pi + 8\sqrt{3} - 2\sqrt{3}\right)\right] = 4\pi - 6\sqrt{3}$.

YOU'VE GOT PROBLEMS

Problem 3: Determine the total area within the graph of the rose $r = 4\sin(3\theta)$.

Example 14: The graphs of the polar curves $r = 4$ and $r = 5 + 2\cos(\theta)$ intersect when $\theta = \frac{2\pi}{3}, \frac{4\pi}{3}$. Find the area of the region that is contained within both of these graphs.

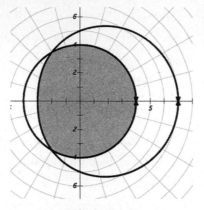

Figure 11.5
The area contained within both $r = 4$ and $r = 5 + 2\cos(\theta)$.

Solution: A complete semicircle makes up part of this region and the area of this semicircle is 8π.

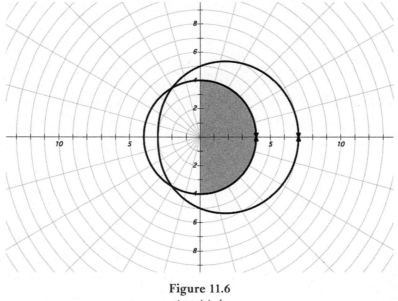

Figure 11.6
A semicircle.

The remaining portion of the region is contained within $r = 5 + 2\cos(\theta)$ and has area

$\frac{1}{2}\int_{2\pi/3}^{4\pi/3}\left(5 + 2\cos(\theta)\right)^2 d\theta$. The total area of the shaded region is $8\pi + \frac{1}{2}\int_{2\pi/3}^{4\pi/3}\left(5 + 2\cos(\theta)\right)^2 d\theta = $

36.953.

The Least You Need to Know

- To find the first derivative of a polar function, you use the equation $\frac{dy}{dx} = \frac{\frac{d(r\sin(\theta))}{d\theta}}{\frac{d(r\cos(\theta))}{d\theta}}$.

- You can write the equation of a line tangent to a polar curve by converting the polar coordinates to rectangular coordinates and also using the slope of the tangent line.

- Find the length of an arc of a polar curve by using the equation
  $L = \int_a^b \sqrt{\left(\frac{dr}{d\theta}\right)^2 + r^2}\, d\theta$.

- To find the area bounded by a polar curve, you use the equation $\int_a^b \frac{1}{2}r^2\, d\theta$.

Introduction to Vectors

In this chapter, we take the time to study vectors. We'll begin with some basic material on vector arithmetic and then work into the calculus of vector functions to examine motion problems.

There is a great deal more to the calculus of vector functions, but that is left to a future course.

In This Chapter

- Understanding the difference between scalars and vectors
- Computing displacement for vector functions
- Calculating velocity for vector functions
- Determining acceleration for vector functions

Scalars and Vectors

Most people use the words *speed* and *velocity* interchangeably, believing that they are describing how fast an object is moving. To mathematicians and physicists, the two terms are related but not the same. Speed does describe the rate at which something is moving, while velocity also gives an indication as to the direction in which the object is moving. Speed is a *scalar* value; it has magnitude but no direction. Velocity is a *vector*; it has both magnitude and direction.

> **DEFINITION**
>
> A **vector** is a quantity that has both magnitude and direction while **scalar** quantities only have magnitude.

Aside from speed versus velocity, we will also discuss distance versus displacement in this chapter. If someone rides 4 miles north and then 3 miles east, how far have they traveled? The answer is 7 miles. This is distance. How far from their starting point are they? The answer is 5 miles (use the Pythagorean Theorem). This is displacement.

Because vectors are directed quantities, they are usually presented graphically with an arrow at one end to indicate direction. If more than one vector is being shown at the same time, the lengths of the segments representing the vectors are drawn proportionally.

As part of a demonstration of vectors to my class, I would take one of the student desks and move it to the front of the room. I would ask a student to come to the front of the room with me and have the student apply a force by pushing the desk from the window side of the room towards the door (see, direction and magnitude). We'd then go to the other side of the room and I'd ask the student to push the desk back toward the position where the desk was before it was pushed. However, just as the student pushed the desk toward the window, I would push the desk at the corner with slightly less force in the direction of the corner of the front wall and the wall containing the windows.

Example 1: Suppose the student pushed the desk with a force of 50 pounds toward the window while I pushed the desk with a force of 40 pounds at angle of 60° to the student's force. What is the net force on the desk?

Solution: A diagram of the action is shown in Figure 12.1.

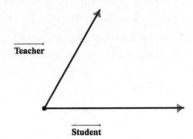

Figure 12.1

A diagram of the two forces acting on the desk (which is represented by a point).

The two forces that are drawn are proportional size to each other and, as you can see, the arrows show the direction in which the force is applied. Notice how each of the forces is labeled. The use of the arrow above the variable name was one of the original methods used to indicate vectors (and is easily used when writing notes on a blackboard).

In print, it is more common to see vectors identified with a bold font or with angled braces such as <a> to represent vectors. The graphical approach for finding the resultant force is called the head-to-tail, or parallelogram, method. Technically, one should draw the first vector, then draw the second vector at the "head" (where the arrow is). The point of the second vector is the "tail." You then connect the tail of the first vector to the head of the second vector (making sure to keep scale and direction accurate). This connection shows the resultant force—with the head of the resultant force being the same point as the head of the second vector. In reality, you are drawing a parallelogram and the resultant force is the diagonal of the parallelogram. This is the addition of vectors.

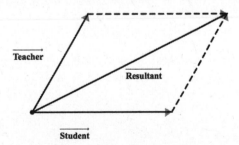

Figure 12.2

The parallelogram method for adding vectors.

You need to go back to your study of trigonometry to determine the length and magnitude of the resultant. Because the angle between the two forces is 60°, the angle opposite the resultant is 120°. Use r to represent the resultant, $r^2 = 40^2 + 50^2 - 2(40)(50)\cos(120)$, so that $r^2 = 6100$ and $r = 78.102$ pounds of force. Use the Law of Sines to determine the angle between the student force and the resultant force is 26.33°.

There is an algebraic way to do this problem as well. The vector **student** has a horizontal component that is 50 units long and a vertical component of 0. To get the horizontal and vertical components for **teacher** we need to use some right triangle trigonometry.

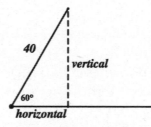

Figure 12.3

Breaking the vector teacher into its horizontal and vertical components.

The horizontal component is $40\cos(60°)$ while the vertical component is $40\sin(60°)$. This makes student = (50,0) and teacher = $\left(20, 20\sqrt{3}\right)$. The sum of these two forces is $\left(50 + 20, 0 + 20\sqrt{3}\right) = \left(70, 20\sqrt{3}\right)$. Pretty easy, wouldn't you say? How long is the resultant? Use the distance formula. (Absolute value is the notation used to determine the length of the resultant.) |resultant| = $\sqrt{70^2 + \left(34.641\right)^2} = 78.102$. (You'll need to be careful when working with approximated values of the components. Let your calculator do all the work.) The angle of the **resultant** from the **student** is $\tan^{-1}\left(\frac{34.641}{70}\right) = 26.33°$.

Example 2: Given the two vectors $a = (4,6)$ and $b = (-2,5)$. Find the length of the resultant for the sum of the two vectors.

Solution: $a + b = (2,11)$. Therefore, $|a + b| = \sqrt{2^2 + 11^2} = \sqrt{125} = 11.18$.

Subtraction of vectors is done exactly as you would think it would be. If $a = (a_1,a_2)$ and $b = (b_1,b_2)$, then $a - b = (a_1 - b_1, a_2 - b_2)$.

There are two forms of multiplication with vectors:

- *Cross product* gives an answer that is a vector.

- *Dot product* (or *inner product*) gives an answer that is a scalar.

The cross product gives a result that is always perpendicular to the vectors being multiplied. That is, if two vectors a and b are in the coordinate plane, then the cross product $a \times b$ will be a vector that is perpendicular to the coordinate plane.

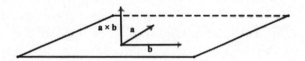

Figure 12.4
The cross product is always perpendicular to the two vectors being multiplied.
The study of the cross product will be covered further in a vector calculus course.

 **DEFINITION**

If $a = (a_1,a_2)$ and $b = (b_1,b_2)$, then $\mathbf{a} \bullet \mathbf{b} = a_1 b_1 + a_2 b_2$ **v.**

Example 3: Given the two vectors $a = (4,6)$ and $b = (-2,5)$. Find the value of $\mathbf{a} \bullet \mathbf{b}$.

Solution: $\mathbf{a} \bullet \mathbf{b} = (4)(-2) + (6)(5) = 22$.

An important formula using the dot product is used to determine the measure of the angle between the two vectors. The formula is $\mathbf{a} \bullet \mathbf{b} = |\mathbf{a}||\mathbf{b}|\cos(\theta)$.

Example 4: Given the two vectors $a = (4,6)$ and $b = (-2,5)$, find the measure of the angle between the two vectors.

Solution: $|a| = \sqrt{4^2 + 6^2} = \sqrt{52}$ and $|b| = \sqrt{(-2)^2 + 5^2} = \sqrt{29}$. Therefore, $\cos(\theta) = \frac{22}{\left(\sqrt{52}\right)\left(\sqrt{29}\right)}$, so $\theta = \cos^{-1}\left(\frac{22}{\left(\sqrt{52}\right)\left(\sqrt{29}\right)}\right) = 55.49°$.

YOU'VE GOT PROBLEMS

Problem 1: Given two vectors $a = (5,9)$ and $b = (-3,10)$, find the value of $a + b$ and the measure of the angle between this sum and a.

Displacement, Velocity, and Acceleration

Because we are able to write vectors in terms of its horizontal and vertical components, we will see a fair amount of the work we did with parametric equations used in the discussion of vector functions. As was the case before, if a function defines the location of an object, then its derivative gives the velocity of the object; the second derivative gives the acceleration of the object.

If the function defines the velocity of an object, then the antiderivative gives the position of the object; the derivative gives the acceleration of the object.

Example 5: A particle moves in the XY plane so that at any time $t \geq 0$ its position (x,y) is given by $x(t) = e^t - e^{-t}$ and $y(t) = e^t + e^{-t}$. Find the velocity vector for any $t \geq 0$.

Solution: The position function is $\mathbf{P} = (x(t),y(t))$, then the velocity function is $(x'(t),y'(t)) = (e^t + e^{-t}, e^t - e^{-t})$.

Example 6: A particle is moving in the coordinate plane. Its position at any time t is given by the vector $(x(t),y(t))$ with $\frac{dx}{dt} = 8t - 6t^2$ and $\frac{dy}{dt} = \ln\left(1 + t^2\right)$. At time $t = 0$, the particle is a t position $(4,-3)$.

1. Determine the rule for the position vector.

2. Determine the velocity and speed of the particle at $t = 1$.

3. Determine the acceleration of the particle at $t = 1$.

Solution:

1. Given $\frac{dx}{dt} = 8t - 6t^2$, $x(t) = \int 8t - 6t^2\ dt\ = 4t^2 - 2t^3 + C$. With $x(0) = 4$, we find that $C = 4$ so $x(t) = 4t^2 - 2t^3 + 4$. $\frac{dy}{dt} = \ln\left(1 + t^2\right)$, so $y(t) = \int \ln\left(1 + t^2\right) dt$. We'll need to use integration by parts to evaluate $\int \ln\left(1 + t^2\right) dt$. Let $u = \ln\left(1 + t^2\right)$ and $dv = dt$, which gives $du = \frac{2t}{1+t^2}$ and $v = t$. Therefore, $\int \ln\left(1 + t^2\right) dt\ =\ t \ln\left(1 + t^2\right) - \int \frac{2t^2}{1+t^2}\ dt\ =$ $t \ln\left(1 + t^2\right) - \int 2 - \frac{2}{1+t^2}\ dt$ (divide $2t^2$ by $1 + t^2$). Consequently, $y(t) = \int \ln\left(1 + t^2\right) dt\ =$ $t \ln\left(1 + t^2\right) - 2t + 2\tan^{-1}(t) + C$. With $y(0) = -3$, $-3 = (0)\ln(1 + 0) - 2(0) + 2\tan^{-1}(0) + C$, so $C = -3$ and $y(t) = t\ln(1 + t^2) - 2t + 2\tan^{-1}(t) - 3$. The position vector for the particle is $(4t^2 - 2t^3 + 4,\ t\ln(1 + t^2) - 2t + 2\tan^{-1}(t) - 3)$.

2. The velocity vector is $(8t - 6t^2, \ln(1 + t^2))$, so the velocity at $t = 1$ is $(2, \ln(2))$. Speed is the absolute value of velocity, so the speed of the particle at $t = 1$ is $\sqrt{4 + \left(\ln(2)\right)^2}$, which is approximately 2.117.

3. The acceleration of the particle is defined by $\left(\frac{d^2x}{dt^2}, \frac{d^2y}{dt^2}\right) = \left(8 - 12t, \frac{2t}{1+t^2}\right)$. At $t = 1$, the acceleration is $(-4, 1)$.

YOU'VE GOT PROBLEMS

Problem 2: An object moves in the plane with velocity vector $(3 + \cos(t), 4 - 2\sin(t))$. Determine the velocity, speed, and acceleration of the particle at $t = \frac{5\pi}{6}$.

Example 7: A particle is moving in the coordinate plane. Its position at any time t is given by the vector $(x(t), y(t))$ with $\frac{dx}{dt} = 8t - 6t^2$ and $\frac{dy}{dt} = \ln\left(1 + t^2\right)$. When is the particle at rest?

Solution: The particle will be at rest when both the horizontal and vertical components of the velocity vector are 0. $8t - 6t^2 = 0$ when $t = 0, \frac{4}{3}$ and $\ln(1 + t^2) = 0$ when $1 + t^2 = e^0$ or when $t = 0$. Therefore, the particle is at rest when $t = 0$.

Example 8: The position of an object as it moves within the coordinate plane is given as $(x(t), y(t))$ with $\frac{dx}{dt} = 8t - 3$ and $\frac{dy}{dt} = 4 - 3t^2$. At $t = 1$, the object is at position $(4,1)$.

1. Determine the speed and acceleration of the object at $t = 1$.

2. Determine the displacement and distance traveled on the interval $1 \leq t \leq 2$.

Solution:

1. The speed of the object is $\sqrt{25 + 1} = \sqrt{26}$. The acceleration vector is $(8, -6t)$. At $t = 1$, this equals $(8, -6)$.

2. The change in the horizontal position of the particle on the interval $1 \leq t \leq 2$ is $\int_1^2 8t - 3 \, dt = 9$, while the change in the vertical position of the particle on the interval $1 \leq t \leq 2$ is $\int_1^2 4 - 3t^2 \, dt = -3$. The displacement of the object is 9 units to the right and 3 units below the original position. The distance traveled is the length of the arc of the path followed. The distance is $\int_1^2 \sqrt{(8t - 3)^2 + (4 - 3t^2)^2} \, dt = 9.664$ units.

YOU'VE GOT PROBLEMS

Problem 3: The position of a particle moving in the coordinate plane for any time on the interval $0 \leq t \leq 2\pi$ is given by $x(t) = 2\sin(t)$ and $y = \cos(2t)$.

(a) Find the velocity and speed of the particle at $t = \frac{2\pi}{3}$.

(b) Determine the time(s) when the particle is at rest.

(c) Determine the displacement of the particle on the interval $0 \leq t \leq 2\pi$. What is the distance traveled by the particle over the same interval?

The Least You Need to Know

- To add and subtract two vectors when the vectors are written in component form, you add/subtract the vectors' corresponding components: $(a_1, b_1) + (a_2, b_2) = (a_1 + b_1, a_2 + b_2)$.

- Compute the dot product of two vectors by $\mathbf{a} \bullet \mathbf{b} = a_1 b_1 + a_2 b_2$.

- Using the dot product, you can find the angle between two vectors.

- To determine the velocity and acceleration vectors from a position vector, you take the first and second derivative of the position vector.

- You can compute the velocity at a defined point of time by evaluating the first derivative of the vector function at that moment in time and the speed of an object by determining the magnitude of the velocity vector.

- To determine the displacement traveled over an interval, subtract the starting vector components from the terminal vector components.

- To determine the distance traveled over an interval, apply the distance formula for the starting and terminal vector values for each interval when the direction motion is the same.

Differential Equations

Many applications in mathematics come from the analysis of the rate at which quantities change. Translated, this means that we have data about the derivative (the rate of change). This is useful information, but it also is often the case that we would like to be able to determine the value of the function whose derivative is known. This is the field of differential equations. We are going to examine a few types of differential equations. If you continue your study of mathematics, you will most likely take a course devoted to just this topic.

In This Chapter

- Solving separable differential equations
- Exploring exponential and logistical growth and decay
- Approximating the value of a function using linear approximations and Euler's method
- Using slope fields to sketch functions
- Solving first order linear differential equations

Separable Differential Equations

The term "separable differential equations" simply means that we are able to manipulate the differential equation so that all terms of one variable are on one side of the equation while all terms of the second variable (and maybe some constants) are on the other side. We can then take the antiderivative of both sides of the equation to determine the function.

Example 1: Solve $\frac{dy}{dx} = xy^2$.

Solution:

1. Gather the terms in y on the left and those in x on the right: $\frac{1}{y^2} dy = x \, dx$.

2. Integrate both sides of the equation. $\int \frac{1}{y^2} dy = \int x \, dx$ gives $\frac{-1}{y} = \frac{1}{2}x^2 + C$. You can solve for y if you so choose. I will not.

Because this solution contains the constant of integration, C, the solution is called a *general solution*. If an initial condition (if a functional value) is known, it is possible to get a *particular solution*.

Example 2: Given $\frac{dy}{dx} = y \cos(x)$ and that $y\left(\frac{\pi}{6}\right) = 2$, express y as a function of x.

Solution:

1. Gather the terms in y on the left and those in x on the right: $\frac{1}{y} dy = \cos(x) \, dx$ and integrate.

2. $\int \frac{1}{y} dy = \int \cos(x) \, dx$ gives $\ln|y| = \sin(x) + C$.

3. Solve this equation for y, $y = e^{\sin(x) + C} = e^c e^{\sin(x)}$.

4. It is traditional to rewrite e^c as a single constant because a constant raised to a constant power is another constant.

5. As a rule, I tend to choose A as this constant, $y = Ae^{\sin(x)}$.

6. Using $y\left(\frac{\pi}{6}\right) = 2$, $y = Ae^{\sin(x)}$ becomes $2 = Ae^{\sin\left(\frac{\pi}{6}\right)} = Ae^{\frac{1}{2}}$ so that $A = \frac{2}{e^{\frac{1}{2}}}$.

7. The particular solution to this differential equation is $y = \frac{2}{e^{\frac{1}{2}}} e^{\sin(x)} = 2e^{\sin(x) - \frac{1}{2}}$.

> ✏️ **CRITICAL POINT**
>
> General solutions to differential equations are left in terms of a constant of integration. Particular solutions use an initial condition to determine a value for the constant of integration.

The differential equation can be given as a verbal description that needs to be translated into an equation.

Example 3: The rate of change of an object traveling in a linear path is directly proportional to the position of the object. At time $t = 0$, the object is at $x = 1$. At time $t = 2$, the object is at $x = 5$. Find the position of the object at $t = 5$.

Solution:

1. The variables are position, x, and time, t.

2. Therefore, the rate of change is given by $\frac{dx}{dt}$.

3. "The position is directly proportional to" tells us that we need a constant of proportion.

4. We'll use k as the constant (because C is the constant of integration and we don't want to get confused).

5. We finally have an equation, $\frac{dx}{dt} = kx$.

6. Gather terms on the appropriate side of the equation. $\frac{1}{x} dx = k\, dtw$ and integrate.
   $\int \frac{1}{x} dx = \int k\, dt$ becomes $\ln|x| = kt + C$.

7. Rewrite the equation with x as a function of t, $x = e^{kt + C} = Ae^{kt}$, with $A = e^c$.

8. Use that $x = 1$ when $t = 0$ to get $1 = Ae^{k(0)}$ to determine that $A = 1$.

9. To find the value of k, use that $x = 5$ when $t = 2$. $5 = e^{k(2)}$ leads to $\ln(5) = 2k$ so that
   $k = \frac{1}{2} \ln(5) = \ln\left(\sqrt{5}\right)$.

10. Therefore, $x = e^{\ln\left(\sqrt{5}\right)t}$ and when $t = 5$, $x = e^{5\ln\left(\sqrt{5}\right)}$ (or 55.902). How's that for moving quickly!

YOU'VE GOT PROBLEMS

Problem 1: Find the particular solution to $\frac{dy}{dx} = \frac{-xy}{\ln(y)}$ ($y > 0$) and the point $(0, e^4)$ is a point on the graph of the function.

Hopefully, by this point in your studies of mathematics, you've either been told or have observed that the rate of change of the function $f(x) = e^x$ at any point of the function is equal to the value of the function at that point. That is, $f'(x) = f(x)$ for all x. Exponential growth and decay problems are a special application of differential equations. They will always fit the description, "the rate of change of the function is proportional to the value of the function." This is exactly what we saw in Example 3.

Example 4: The rate of change of North Americans who drink carb-free energy drinks is proportional to the number of people who drink them. In the year 2000, consumers drank 180,000 gallons of carb-free energy drinks, and in 2010, that number rose to 1.1 million gallons. If growth continues in the same manner, predict the volume of sales for 2020.

Solution:

1. Use the year 2000 as time $t = 0$. The differential equation $\frac{dV}{dt} = kV$ leads to the equation $V = Ae^{kt}$.

2. The initial data gives $180000 = Ae^{k(0)}$ so that $A = 180000$.

3. The data from 2010 gives the equation $1100000 = 180000e^{10k}$.

4. Solve for k: $\frac{110}{18} = e^{10k}$ so that $k = \frac{1}{10}\ln\left(\frac{110}{18}\right)$ (or 0.181011). (There are more computations to be done, so it is best to store the exact answers into a variable on your calculator.)

5. The predicted value for 2020 is $V = 180000e^{\left(\frac{1}{10}\ln\left(\frac{110}{18}\right)\right)(20)}$, which is approximately 6.7 million gallons.

Exponential decay problems will look exactly the same with the exception that the values will decrease as time goes forward and the constant of proportionality will be negative.

Newton's Law of Heating and Cooling says that the rate of change of the temperature of an object is proportional to the difference of the environment and the temperature of the object. For most of us, the "environment" is usually the oven or the freezer.

Example 5: When cooking a turkey for Thanksgiving, Diane will take the turkey out of the refrigerator, quickly dress and prep it, and put it into an oven that has been preheated to 325°F. She will cook the turkey until it reaches a temperature of 165°F. She uses a meat thermometer to determine the temperature of the turkey one hour after she put it in the oven and finds that the temperature is 80°F. How much time is needed for the turkey to cook?

Solution: The most important problem here is what do we name the variable. Do we use T for turkey or will that confuse us with t for time? We can use U, the second letter in turkey, or y, the last letter in turkey, but I think I'll use B because everyone always asks when the bird will be done. Okay, now that we have that issue settled, we can get down to business.

1. The temperature of the environment is 325°F and the initial temperature of the bird is about 45°F. (The turkey was in the refrigerator at a temperature of 40°F and has been out of the refrigerator for less than 30 minutes, so it didn't warm up too much. An estimate of 45°F is probably a bit high, but it will work for now.)

2. The differential equation is $\frac{dB}{dt} = k\left(325 - B\right)$. This becomes $\frac{1}{325 - B}\, dB = k\, dt$.

3. Integrate and solve: $\int \frac{1}{325 - B}\, dB = \int k\, dt$ becomes $-\ln(325 - B) = kt + C$. Multiply through by -1 to get $n(325 - B) = kt + C$, letting both constants k and C absorb the negative sign. We then get $B = 325 - Ae^{kt}$, with $A = e^c$. The initial temperature of the bird is 45°, so 45 $= 325 - Ae^{k(0)}$ and $A = 280$.

4. Use the temperature taken at the 1-hour mark to determine the value of k. $80 = 325 - 280^{k(1)}$ gives $k = \ln\left(\frac{7}{8}\right)$.

5. Finally, find the amount of time it will take for the turkey to reach the desired temperature. $165 = 325 - 280e^{\ln\left(\frac{7}{8}\right)t}$ gives $t = 4.2$. The bird will be ready approximately 4 hours and 10 minutes after it was placed in the oven.

Another type of special application is the logistical equation. Although they initially look like exponential functions, the logistical equation eventually levels off.

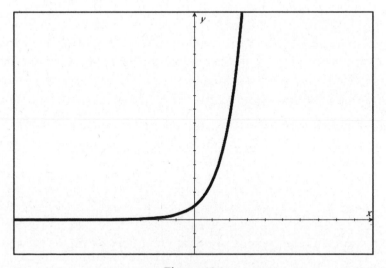

Figure 13.1
Example of exponential growth.

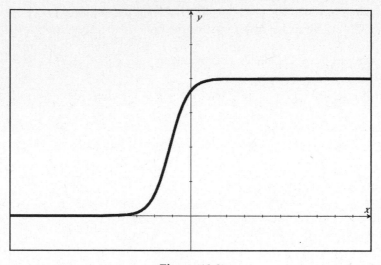

Figure 13.2
Example of logistical growth.

In logistical growth problems, there is always a capacity that caps the amount of change. For example, if one is measuring the rate with which a disease is spreading through a population, the size of the population is the largest number of people who can have the disease. In population growth, bacteria are limited by the size of the petri dish in which they are being cultivated. Human population is limited by the ability to feed itself.

Example 6: A 2,500-gallon aquarium can support no more than 125 guppies. Ten guppies are introduced into the aquarium. Assume that the rate of guppy population growth is directly proportional to the population y and the limiting factor $125 - y$ at any time t (weeks) with proportionality factor $k = 0.0025$.

(a) Determine the guppy population $y(t)$ as an explicit function of time t.

(b) What values of y and t make sense in the population problem?

(c) When should the guppy population reach the aquarium's capacity?

Solution:

(a)

1. The differential equation for this problem is $\frac{dy}{dt} = 0.0025y(125 - y)$, so that the integration problem becomes $\int \frac{1}{y(125 - y)} \, dy = \int 0.0025 \, dt$.

2. The integral on the right-hand side of the equation is simple enough,

    $\int 0.0025\, dt = 0.0025t + C$. However, the integral on the left-hand side is going to require

    that we apply the partial fraction technique.

    $\frac{1}{y(125-y)} = \frac{A}{y} + \frac{B}{125-y}$ and $1 = A(125-y) + By$

3. Set $y = 0$ to get $A = \frac{1}{125}$; set $y = 125$ to get $B = \frac{1}{125}$. This changes $\int \frac{1}{y(125-y)}\, dy$ to

    $\frac{1}{125} \int \frac{1}{y} + \frac{1}{125-y}\, dy = \frac{1}{125}\left(\ln|y| - \ln|125 - y|\right) = \frac{1}{125} \ln\left|\frac{y}{125-y}\right|$.

4. Now we know $\int \frac{1}{y(125-y)}\, dy = \int 0.0025\, dt$ leads to $\frac{1}{125}\ln\left|\frac{y}{125-y}\right| = 0.0025t + C$. Multiply

    by 125, $\ln\left|\frac{y}{125-y}\right| = 0.3125t + C$. (125 times C is still a constant and I am just going to let

    C absorb the 125.)

5. Rewrite this as an exponential equation, $\frac{y}{125-y} = Ae^{0.3125t}$ (once again, $A = e^c$).

6. Multiply through by the common denominator.

    $y = (125 - y)Ae^{0.3125t} = 125Ae^{0.3125t} - yAe^{0.3125t}$

7. Gather the terms in y:

    $y + yAe^{0.3125t} = 125Ae^{0.3125t}$

8. Factor and solve for y:

    $y = \frac{125Ae^{0.3125t}}{1 + Ae^{0.3125t}}$

9. We can now use the fact that the initial population was 10 guppies. $10 = \frac{125Ae^{0.3125(0)}}{1 + Ae^{0.3125(0)}}$ so

    that $10 = \frac{125A}{1+A}$ and $10 + 10A = 125A$, or that $A = \frac{10}{115} = \frac{2}{23}$.

10. Finally we have $y = \frac{125\left(\frac{2}{23}\right)e^{0.3125t}}{1 + \left(\frac{2}{23}\right)e^{0.3125t}} = \frac{250e^{0.3125t}}{23 + 2e^{0.3125t}}$.

(b)

1. Time begins when the guppies are put into the tank, so the values of t that make sense
    are $t \geq 0$.

2. The number of guppies starts with 10.

3. If they survive, the total will never exceed 125, which is the capacity of the tank.

4. Time is a continuous variable while the number of guppies is discrete.

(c)

Never. If we solve the equation $125 = \frac{250e^{0.3125t}}{23 + 2e^{0.3125t}}$ we get $125(23 + 2e^{0.3125t}) = 250e^{0.3125t}$, so that $125(23) = 0$. The limiting value for this function is 125.

BE AWARE

With exponential growth, the rate of change is proportional to the amount present. With logistical growth, the rate of change is proportional to the product of the amounts present and not present.

YOU'VE GOT PROBLEMS

Problem 2: When a water-cooled nuclear power plant is operating, oxygen in the water is transmuted to an isotope of nitrogen. After the reactor is shut down, the radiation from this nitrogen decreases in such a way that the rate of change of the radiation level is proportional to the radiation level.

(a) Write a differential equation that expresses the rate of change of the radiation level in terms of the radiation level. Solve the equation and express the radiation level as a function of time.

(b) Suppose that when the reactor is first shut down, the radiation level is 4×10^{18} units. After 90 seconds, the level has dropped to 5×10^{14} units. Write the particular equation.

(c) It is safe to enter the reactor compartment when the radiation level has dropped to 8×10^{-4} units. When will it be safe to enter the reactor compartment?

Linear Approximations

In the days before calculators, linear approximation was a worthwhile (if not important) topic for its practical applications. The basic idea of linear approximations is that if you "zoom in" on any graph with sufficient magnitude, the graph will look linear. Here is a picture of the graph of $y = \sin(x)$ on the interval $[-0.01, 0.01]$.

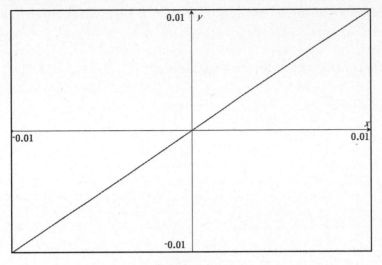

Figure 13.3
The graph of y = sin(x) on the interval [−0.01,0.01].

1. Why was the topic worthwhile? Suppose we need an estimate for $\sqrt[3]{67}$.

2. We know that $\sqrt[3]{64} = 4$. We'd use the line tangent to the graph of $y = \sqrt[3]{x}$ to get the estimate.

3. The slope of the tangent line to $y = \sqrt[3]{x}$ at $x = 64$ is $\frac{1}{3}(64)^{-\frac{2}{3}} = \frac{1}{48}$.

4. The equation of the tangent line at the point when $x = 64$ is $y - 4 = \frac{1}{48}(x - 64)$.

5. We want $\sqrt[3]{67}$, so we substitute $x = 67$ to get $y - 4 = \frac{1}{48}(67 - 64)$, so that $y = 4\frac{1}{16} = 4.0625$.

Compare that to the value on your calculator, $\sqrt[3]{67} = 4.06155$. (The result found by linear approximation is less than 0.03 percent in error.)

Of course, we would often get fractional results that were more challenging to convert to a decimal than $\frac{1}{16}$, but that was not as significant as it might seem. If calculators can do the work more quickly and efficiently than linear approximation, why raise the issue? There are two reasons:

- Linear approximation is the preliminary step to doing more involved approximations. Later in this book, we will look at power series as a means of approximating more complicated functions.

- There is no cubed-root function on your calculator. That button activates a power series to compute the value in question. The same can be said for all the trigonometric, logarithmic, and exponential functions on your calculator. As my Linear Algebra professor at Manhattan College, Sylvester Tuohy, used to say, "This is good stuff!"

Example 7: Estimate $\sin^{-1}\left(\frac{3}{5}\right)$ to 4 decimal places.

Solution:

1. We know $\sin^{-1}\left(\frac{1}{2}\right)=\frac{\pi}{6}$, so we can work from this point as it is close to our desired point.

2. The derivative of $y = \sin^{-1}(x)$ is $\frac{1}{\sqrt{1-x^2}}$ and the value of this derivative at $x=\frac{1}{2}$ is

$$\frac{1}{\sqrt{1-\left(\frac{1}{2}\right)^2}}=\frac{1}{\sqrt{1-\frac{1}{4}}}=\frac{1}{\sqrt{\frac{3}{4}}}=\frac{2}{\sqrt{3}}=\frac{2\sqrt{3}}{3}.$$

3. The equation of the tangent line at $\left(\frac{1}{2},\frac{\pi}{6}\right)$ is $y-\frac{\pi}{6}=\frac{2\sqrt{3}}{3}\left(x-\frac{1}{2}\right)$.

4. At $x=\frac{3}{5}$, $y-\frac{\pi}{6}=\frac{2\sqrt{3}}{3}\left(\frac{3}{5}-\frac{1}{2}\right)$. (Okay, at this stage, you have to be asking yourself, *Why are you doing this to me?* Because it's fun. That sounds a bit snobbish, but this material is still in some books. So let's say we just slog through this problem and I promise not to raise the issue again.)

5. We know $\pi = 3.14156$ (approximately), so $\frac{\pi}{6}=0.5236$ and $\sqrt{3}=1.732$. (Did you know that George Washington was born in 1732? I saw this when I was a kid and have never forgotten it and, you know, whenever I mention it I turn around to see if someone is coming to drag me off the street.) Where was I? Oh, yes.

6. $\frac{2\sqrt{3}}{3}=2(0.5774)=1.1548$, so $y-\frac{\pi}{6}=\frac{2\sqrt{3}}{3}\left(\frac{3}{5}-\frac{1}{2}\right)$ becomes $y-0.5326 = 1.1548(.1)$ and $y = 0.64808$. (Not too bad; off by less than 1 percent.)

Euler's Method

One of the problems raised by the process for linear approximations is that you are not always guaranteed that the value whose functional value you seek is near a value whose functional value you know. Euler's Method attempts to get a more accurate approximation by moving from a known point to the unknown point in small steps and, in essence, perform linear approximations on a repeated basis. It is anticipated that you will use technology in this process.

CRITICAL POINT

Leonhard Euler, who was voted the third most influential mathematician from the second millennium as part of the craze that was Y2K, was a prolific writer of mathematics. His work involved most of the fields in mathematics and the number e is named after him.

Example 8: Let $f(x)$ be a function with the properties $f'(x) = 4f(x)$ and $f(5) = 3$. Approximate $f(5.4)$ using increments of 0.1.

Solution: I will tell you that the linear approximation method gives the result 7.8. (Check it out if you want.) I will write out the solution in all its glorious details at first and then show you the tabular method that you will use for all the other problems:

1. An increment of 0.1 takes us from 5 to 5.1. The first step takes us from 3 to
   $y = 3 + 12(5.1 - 5) = 4.2$.

2. We're now at the point $(5.1, 4.2)$. $f'(5.1) = 4(4.2) = 16.8$. Our next estimate is
   $y = 4.2 + 16.8(5.2 - 5.1) = 5.88$.

3. We are now at the point $(5.2, 5.88)$. $f'(5.2) = 4(5.88) = 23.52$. Our next estimate is
   $y = 5.88 + 23.52(5.3 - 5.2) = 8.232$.

4. From $(5.3, 8.232)$ we have $f'(5.3) = 4(8.232) = 32.928$. Our estimate for $f(5.4)$ is
   $y = 8.232 + 32.928(5.4 - 5.3) = 11.5248$.

Working from a table, we need:

- The starting value of x and y

- The amount of the increment, dx

- To compute $\frac{dy}{dx}$

- To compute dy as the product of dx and $\frac{dy}{dx}$

Here is the table for the same problem.

x	y	dx	$\frac{dy}{dx}$	dy
5	3	0.1	12	1.2
5.1	4.2	0.1	16.8	1.68
5.2	5.88	0.1	23.52	2.352
5.3	8.232	0.1	32.928	3.2928
5.4	11.5248			

Example 9: Given the differential equation $\frac{dy}{dx} = \frac{-x^2 y}{2}$ with the initial condition that f(2) = 5, use Euler's Method to approximate f(2.1) using increments of 0.05.

Solution: Let's go directly to the table method for this problem.

x	y	dx	$\frac{dy}{dx}$	dy
2	5	0.05	−10	−0.5
2.05	4.5	0.05	−9.45563	−0.472781
2.1	4.02722			

YOU'VE GOT PROBLEMS

Problem 3: Given the differential equation $\frac{dy}{dx} = 2y(10 - y)$ with the condition f(3) = 2 is a point on the original graph. Use Euler's Method to approximate f(3.2) using increments of 0.1.

Slope Fields

Slope fields are a visual representation of a differential equation. It is traditional to use lattice points (points whose coordinates are integers) to compute the slope of the tangent line to the function using the differential equation. Draw a small line segment with that slope at that point. Once an initial condition for the function is known, it is possible to sketch a graph of the function by following the tangents.

DEFINITION

> A **slope field** shows the slopes of tangent lines for a family of solutions to a given differential equation.

The slope field for the differential equation $\frac{dy}{dx} = \frac{1}{2}x^2$ is shown.

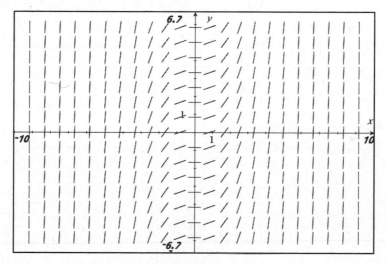

Figure 13.4
Slope field for $\frac{dy}{dx} = \frac{1}{2}x^2$.

We know that the solution to this differential equation is $y = \frac{1}{6}x^3 + C$. If we set the initial condition to be at the point (0,0), the graph would be:

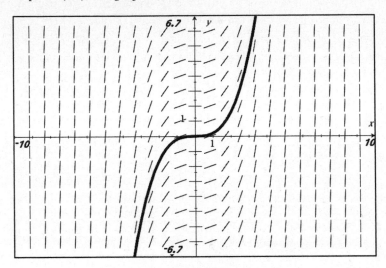

Figure 13.5

The graph of $y = \frac{1}{6}x^3$ on the slope field for $\frac{dy}{dx}$.

If the initial condition was the point (1,2), the graph would be:

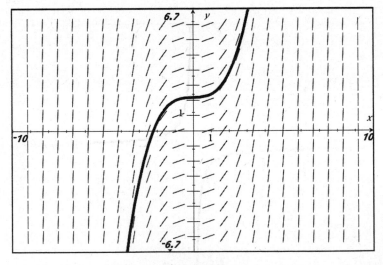

Figure 13.6

The graph of $y = \frac{1}{6}x^3 + \frac{11}{6}$ on the slope field for $\frac{dy}{dx}$.

Example 10: Given the slope field for the differential equation $\frac{dy}{dx} = \frac{-y}{x}$, sketch a solution to this differential equation using the initial value (1,1).

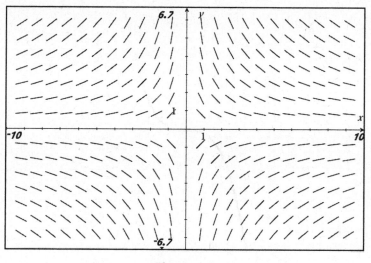

Figure 13.7

Slope field for $\frac{dy}{dx} = \frac{-y}{x}$.

Solution: Work from the point (1,1) to sketch a branch of the function. Use symmetry to sketch the other branch of the hyperbola.

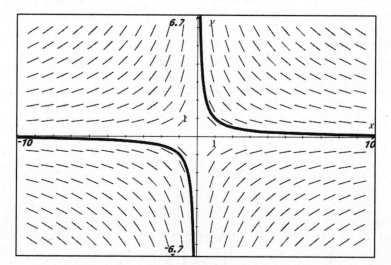

Figure 13.8

Graph of $y = \frac{1}{x}$ on the slope field for $\frac{dy}{dx} = \frac{-y}{x}$.

Example 11: Draw the slope field for the differential equation $\frac{dy}{dx} = xy$ on the graph with the lattice points.

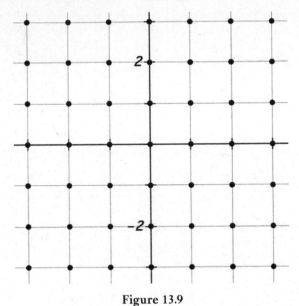

Figure 13.9
Lattice points on a coordinate grid.

Solution: The grid contains the points $-3 \le x \le 3$ and $-3 \le y \le 3$. Let's make a table to compute the slope for each of these lattice points.

	−3	−2	−1	0	1	2	3
−3	9	6	3	0	−3	−6	−9
−2	6	4	2	0	−2	−4	−6
−1	3	2	1	0	−1	−2	−3
0	0	0	0	0	0	0	0
1	−3	−2	−1	0	1	2	3
2	−6	−4	−2	0	2	4	6
3	−9	−6	−3	0	3	6	9

Draw a segment with the given slope at the appropriate lattice points.

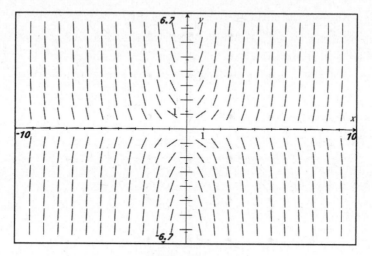

Figure 13.10

Slope field for $\frac{dy}{dx} = xy$.

YOU'VE GOT PROBLEMS

Problem 4: The slope field for the differential equation $\frac{dy}{dx} = 2y$ is shown. Sketch the graph of the particular solution containing the point (1,1). differential equation

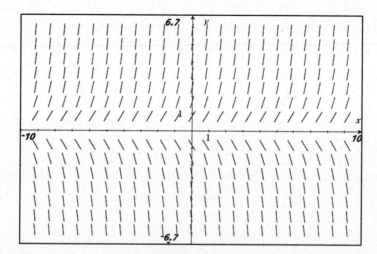

Figure 13.11

Slope field for $\frac{dy}{dx} = 2y$.

First Order Linear Differential Equations

First order linear differential equations are always of the form $\frac{dy}{dt} + P(t)y = Q(t)$. (Differential equations can take other forms, but they would not be classified as first order linear differential equations. Remember, there is an entire course dedicated to differential equations.) The process for solving these equations seems a bit strange (well, that's because it *is* strange). The process uses an expression that is called an *integration factor,* $u(x) = e^{\int P(t)\,dt}$. Multiply through both sides of the equation by this integration factor to get:

$$e^{\int P(t)\,dt}\left(\frac{dy}{dt} + P(t)y\right) = e^{\int P(t)\,dt}\,Q(t)$$

This becomes:

$$e^{\int P(t)\,dt}\,\frac{dy}{dt} + P(t)e^{\int P(t)\,dt}\,y = e^{\int P(t)\,dt}\,Q(t)$$

The genius of this method is that the left-hand side of this equation is the derivative of the product $e^{\int P(t)\,dt}\,y$. Be sure you understand what this means,

$$\int\left(e^{\int P(t)\,dt}\,\frac{dy}{dt} + P(t)e^{\int P(t)\,dt}\,y\right)dt = e^{\int P(t)\,dt}\,y \text{ (with the constant of integration to be}$$

named later). The notation $e^{\int P(t)\,dt}\,y$ might look too intimidating. Rather than write

$$\int\left(e^{\int P(t)\,dt}\,\frac{dy}{dt} + P(t)e^{\int P(t)\,dt}\,y\right)dt = e^{\int P(t)\,dt}\,y, \text{ I should write } \int\left(u(t)\frac{dy}{dt} + P(t)u(t)y\right)dt = u(t)y$$

because that looks better. What do you think about that?

Back to work. Integrating both sides of the equation $u(t)\frac{dy}{dt} + P(t)u(t)y = u(t)\,Q(t)$ gives

$$u(t)y = \int u(t)\,Q(t)\,dt + C \text{ so that } y = \frac{\int u(t)\,Q(t)\,dt + C}{u(t)}.$$

Let's see if this actually works. (Well, of course it works, but we need to get the hang of it.)

Example 12: Solve $\frac{dy}{dx} + 4xy = x$.

Solution:

1. The problem fits the form of a first order linear differential equation. The integration factor is $e^{\int 4x\,dx} = e^{2x^2}$.

2. Multiplying through both sides of the equation gives $e^{2x^2}\frac{dy}{dx} + 4xe^{2x^2}y = xe^{2x^2}$.

3. Integrate both sides of the equation:

$$\int \left(e^{2x^2} \tfrac{dy}{dx} + 4xe^{2x^2} y \right) dx = \int xe^{2x^2} dx$$

4. Simplify both sides of the equation:

$$e^{2x^2} y = \int xe^{2x^2} dx = \tfrac{1}{4} e^{2x^2} + C$$

5. Solve for y:

$$y = \frac{\tfrac{1}{4} e^{2x^2} + C}{e^{2x^2}} = \tfrac{1}{4} + Ce^{-2x^2}$$

That wasn't too bad was it? Ready for another one?

Example 13: Solve $x \tfrac{dy}{dx} + y = x^3 + 4$ for $x > 0$.

Solution:

1. Don't get fooled—this doesn't quite fit the pattern for the first order linear differential equation.

2. We must first divide by the expression in front of the derivative.

3. $x \tfrac{dy}{dx} + y = x^3 + 4$ must be rewritten as $\tfrac{dy}{dx} + \tfrac{y}{x} = x^2 + \tfrac{4}{x}$.

4. The integration factor is $u(x) = e^{\int \frac{1}{x} dx} = e^{\ln|x|} = x$.

5. Multiply by the integration factor:

$$x \tfrac{dy}{dx} + y = x^3 + 4$$

(Look at that—we're back where we started with this one.)

6. Integrate both sides of the equation:

$$\int \left(x \tfrac{dy}{dx} + y \right) dx = \int x^3 + 4 \, dx$$

7. Simplify both sides of the equation:

$$xy = \int x^3 + 4 \, dx = \tfrac{1}{4} x^4 + 4x + C$$

8. Solve for y:

$$y = \tfrac{1}{4} x^3 + 4 + \tfrac{C}{x}$$

Example 14: Solve $\cos^2(x)\frac{dy}{dx}+\sin(2x)y=2\cos^3(x)\sin(x)-1$ with $0<x<\frac{\pi}{2}$.

Solution:

1. We're old pros at this now. Get the equation into the format of a first order linear differential equation by dividing both sides of the equation by $\cos^2(x)$,

    $\cos^2(x)\frac{dy}{dx}+\sin(2x)y=2\cos^3(x)\sin(x)-1$ becomes $\frac{dy}{dx}+\frac{\sin(2x)}{\cos^2(x)}y=\frac{2\cos^3(x)\sin(x)-1}{\cos^2(x)}$.

2. Simplify the term $\frac{\sin(2x)}{\cos^2(x)}=\frac{2\sin(x)\cos(x)}{\cos^2(x)}=2\tan(x)$.

3. The problem is now $\frac{dy}{dx}+2\tan(x)y=2\cos(x)\sin(x)-\sec^2(x)$.

4. The integration factor for this problem then is $u(x)=e^{\int 2\tan(x)\,dx}=e^{2\ln|\sec(x)|}=\sec^2(x)$.

5. Multiply by the integration factor:

    $\sec^2(x)\frac{dy}{dx}+2\tan(x)\sec^2(x)y=2\cos(x)\sin(x)\sec^2(x)-\sec^4(x)$

6. Integrate both sides of the equation:

    $\int\left(\sec^2(x)\frac{dy}{dx}+2\tan(x)\sec^2(x)y\right)dx=\int 2\cos(x)\sin(x)\sec^2(x)-\sec^4(x)\,dx$

7. Simplify both sides of the equation:

    $\sec^2(x)y=\int 2\tan(x)-\sec^4(x)\,dx$ (Remember: $\cos(x)\sec(x)=1$.)

8. Unlike the other problems we've done where we just wrote the integral out, we need to take time to look at the integral $\int\sec^4(x)\,dx$. We'll rewrite it as

    $\int\sec^2(x)\left(1+\tan^2(x)\right)dx=\int\sec^2(x)+\sec^2(x)\tan^2(x)\,dx$.

9. Continuing:

    $\sec^2(x)y=2\ln\left|\sec(x)\right|-\tan(x)+\frac{1}{3}\tan^3(x)+C$

10. Solve for y:

    $y=\frac{2\ln|\sec(x)|-\tan(x)-\frac{1}{3}\tan^3(x)+C}{\sec^2(x)}=2\cos^2(x)\ln\left|\sec(x)\right|-\tan(x)\cos^2(x)-\frac{1}{3}\tan^3(x)\cos^2(x)+C\cos^2(x)$

    $=2\cos^2(x)\ln\left|\sec(x)\right|-\sin(x)\cos(x)-\frac{1}{3}\tan(x)\sin^2(x)+C\cos^2(x)$

YOU'VE GOT PROBLEMS

Problem 5: Solve $\frac{dy}{dx}+2y=x+3$.

The Least You Need to Know

- To solve a separable differential equation, you must gather terms of the same variable on one side of the equation.

- Solve a first order linear differentiable equation by multiplying both sides of the equation by the integration factor.

- To distinguish between exponential growth and logistical growth, examine the quantity to which the rate of change is proportional.

- Apply Euler's Method to a differentiable equation to get an approximation of a function value.

- Sketch a graph given a slope field and an initial point by evaluating the derivative at each of the lattice points on the grid.

CHAPTER 14

Infinite Sequences

The study of sequences at this level of mathematics serves primarily as a precursor for the study of series. In this chapter, we study sequences that are defined by formulas as well as those defined by a pattern.

In This Chapter

- Examining the convergence and divergence of sequences
- Applying the Squeeze Theorem to determine convergence
- Examining increasing, decreasing, and monotonic sequences

Convergence and Divergence of Sequences

A *sequence* is a listing of numbers. In our discussion, that list will be infinitely long. Although there are cases where the entries in the list have no rhyme or reason, we will limit our discussion to sequences that have a defined pattern. Defined patterns can come as implicit definition or a recursive definition. The terms of the sequence are usually represented in terms of the letter a with subscripts used to represent the position in the sequence. That is, a sequence is represented as $a_1, a_2, a_3, \dots, a_n$.

The sequence can be symbolized as $\{a_n\}$. This means the sequence $\{2n - 1\}$ would be the sequence of odd counting numbers, 1, 3, 5, 7, ...

A sequence can be defined recursively, also known as a *recursively defined sequence.*

DEFINITION

A **sequence** defined implicitly is the output of a function f(x) in which the domain is the set of positive integers. A **recursively defined sequence** has a defined initial value, or values, and for some value c where $n > c$, the term a_n is written in terms of previous values of a.

Example 1: Write the first five terms of the sequence defined by $a_1 = 4$ and $a_n = a_{n-1} + 3$.

Solution: We have $a_1 = 4$, so $a_2 = a_1 + 3 = 4 + 3 = 7$, $a_3 = a_2 + 3 = 7 + 3 = 10$, $a_4 = a_3 + 3 = 10 + 3 = 13$, and $a_5 = a_4 + 3 = 13 + 3 = 16$. The first five terms are 4, 7, 10, 13, 16. You might recognize this as an arithmetic sequence, which is a sequence in which the difference between successive terms is a constant. This sequence can also be defined as $\{3n + 1\}$.

Example 2: Write the first five terms of the sequence defined by $a_1 = 4$ and $a_n = 3a_{n-1}$.

Solution: We have $a_1 = 4$, so $a_2 = 3a_1 = 12$, $a_3 = 3a_2 = 36$, $a_4 = 3a_3 = 108$, and $a_5 = 3a_4 = 324$. The first five terms are 4, 12, 36, 108, 324. You might recognize this as a geometric sequence, which is a sequence in which the ratio of successive terms is a constant. This sequence can also be defined as $\{3(4)^{n-1}\}$.

Example 3: Write the first seven terms of the sequence defined by $a_1 = 1$, $a_2 = 1$, and $a_n = a_{n-1} + a_{n-2}$ for $n \geq 3$.

Solution:

1. The first two terms are already defined, so let's look at the third term: $a_3 = a_{3-1} + a_{3-2} = a_2 + a_1 = 1 + 1 = 2$.

2. The fourth term is $a_4 = a_{4-1} + a_{4-2} = a_3 + a_2 = 2 + 1 = 3$, while the fifth term is $a_5 = a_4 + a_3 = 3 + 2 = 5$.

3. The sixth and seventh terms are $a_6 = a_5 + a_4 = 5 + 3 = 8$ and $a_7 = a_6 + a_5 = 8 + 5 = 13$.

4. The first seven terms are 1, 1, 2, 3, 5, 8, and 13. This is the famous Fibonacci Sequence.

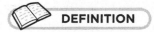

CRITICAL POINT

> Fibonacci was a thirteenth century mathematician who is famous for using the Hindu Arabic numerals (0, 1, 2, 3, ...) in his book *Liber Abaci*. His famous sequence was an attempt to predict the size of rabbit populations (which he raised) under ideal conditions.

The limit of a sequence, L, is the value that the terms a_n grow to as n gets very large, if that value exists. That is, $\lim\limits_{n \to \infty} a_n = L$.

Example 4: Determine the limiting value for $\left\{ \frac{n}{n+1} \right\}$.

Solution:

1. To determine the value of $\lim\limits_{n \to \infty} \frac{n}{n+1}$, we can use the definition for the implicitly defined sequences.

2. $\left\{ \frac{n}{n+1} \right\}$ is generated by the function $f(x) = \frac{x}{x+1}$ and $\lim\limits_{x \to \infty} f(x) = 1$ when we apply L'Hopital's Rule.

3. Therefore, $L = 1$.

Because the limit is a finite number, we say the sequence *converges*.

DEFINITION

> If the $\lim\limits_{n \to \infty} a_n$ exists and is finite, we say the sequence $\left\{ a_n \right\}$ is **convergent.**
> If the limit fails to exist or is infinity, we say that the sequence *diverges*.

Example 5: Determine if the sequence $\left\{\frac{\ln(n)}{n}\right\}$ converges or diverges.

Solution:

1. We need to evaluate $\lim\limits_{n \to \infty} \frac{\ln(n)}{n}$.

2. To do so, apply L'Hopital's Rule to the function $f(x) = \frac{\ln(x)}{x}$, $\lim\limits_{x \to \infty} \frac{\ln(x)}{x} = \lim\limits_{x \to \infty} \frac{\frac{1}{x}}{1} = 0$.

3. Therefore, $\lim\limits_{n \to \infty} \frac{\ln(n)}{n} = 0$.

4. This means $\left\{\frac{\ln(n)}{n}\right\}$ converges.

Example 6: Determine if the sequence $\left\{\frac{e^n}{n}\right\}$ converges or diverges.

Solution:

1. We need to evaluate $\lim\limits_{n \to \infty} \frac{e^n}{n}$.

2. Apply L'Hopital's Rule to the function $f(x) = \frac{e^x}{x}$, $\lim\limits_{x \to \infty} \frac{e^x}{x} = \lim\limits_{x \to \infty} \frac{e^x}{1} = \infty$.

3. Therefore, $\left\{\frac{e^n}{n}\right\}$ diverges.

Example 7: Given the sequence defined recursively as $a_1 = 1$ and $a_{n+1} = \sqrt{12 + a_n}$, determine the limit for this sequence as n goes to infinity.

Solution:

1. The key to solving this problem is to realize that $\lim\limits_{n \to \infty} a_{n+1} = L$ and $\lim\limits_{n \to \infty} a_n = L$.

2. Therefore, $L = \sqrt{12 + L}$, so that $L^2 = 12 + L$ or that $L^2 - L - 12 = 0$, $(L-4)(L+3) = 0$, so $L = 4, -3$.

3. Because the limit cannot be negative in this case, the range of the square root function is non-negative, $L = 4$.

CRITICAL POINT

If $\{a_n\}$ and $\{b_n\}$ both converge, then so does $\{a_n + b_n\}$, $\{a_n - b_n\}$, $\{a_n \times b_n\}$. If $\lim_{n \to \infty} b_n$ does not equal 0, $\left\{\frac{a_n}{b_n}\right\}$ also converges. If k is a constant, then $\{ka_n\}$ also converges. If $\{a_n\}$ and $\{b_n\}$ both diverge, then so does $\{a_n + b_n\}$, $\{a_n - b_n\}$, $\{a_n \times b_n\}$. If k is a constant, then $\{ka_n\}$ also diverges. If either $\{a_n\}$ or $\{b_n\}$ diverges, then so does, $\{a_n + b_n\}$, $\{a_n - b_n\}$, $\{a_n \times b_n\}$.

Example 8: Given the sequences $\{3^{-n}\}$ and $\{n^2 4^{-n}\}$, determine if the sequence $\{3^{-n} + n^2 4^{-n}\}$ converges or diverges.

Solution:

1. The $\lim_{n \to \infty} 3^{-n} = 0$, but we need to determine if the $\lim_{n \to \infty} n^2 4^{-n}$ exists and is finite.

2. Use the function $f(x) = \frac{x^2}{4^x}$ and L'Hopital's Rule. $\lim_{x \to \infty} \frac{x^2}{4^x} = \lim_{x \to \infty} \frac{2x}{4^x \ln(4)} = \lim_{x \to \infty} \frac{2}{4^x (\ln(4))^2} = 0$. (Recall that the derivative of b^x is $b^x \ln(b)$.)

3. Each of the sequences converges, so the sum of the sequences does, too.

YOU'VE GOT PROBLEMS

Problem 1: Determine if $\left\{\frac{n^2 + n + 10000}{2n^2 - 1}\right\}$ converges or diverges.

Squeeze Theorem

We will run into examples whose limit is a bit more challenging to evaluate then simply using L'Hopital's Rule. The Squeeze Theorem provides us with another tool to attack problems.

CRITICAL POINT

If $a_n \le b_n \le c_n$ for $n \ge k$, in which k is a constant, $\lim_{n \to \infty} a_n = L$, and $\lim_{n \to \infty} c_n = L$ then $\lim_{n \to \infty} b_n = L$.

The Squeeze Theorem requires that we are able to find two known sequences that have the same limit and are bound to the unknown function. We won't be able to do this with all sequences, but there are a few patterns to which we can apply the theorem.

Example 9: Determine if the sequence $\left\{\frac{\cos(n)}{n}\right\}$ is convergent.

Solution:

1. You might be tempted to think that you can use L'Hopital's Rule here, but that is not so. The $\lim\limits_{x\to\infty} \cos(x)$ cannot be determined because it oscillates between -1 and 1. Wait a minute! That's the answer.

2. We know that $\frac{-1}{n} \le \frac{\cos(n)}{n} \le \frac{1}{n}$, and we know that $\lim\limits_{n\to\infty} \frac{-1}{n} = \lim\limits_{n\to\infty} \frac{1}{n} = 0$.

3. The Squeeze Theorem tells us that $\lim\limits_{n\to\infty} \frac{\cos(n)}{n} = 0$.

4. Therefore, $\left\{\frac{\cos(n)}{n}\right\}$ is convergent.

Example 10: Determine if the sequence $\left\{\frac{n!}{n^n}\right\}$ converges. ($n!$, read as n factorial, is the product of the first n positive integers.)

Solution:

1. We cannot use L'Hopital's Rule on this problem because $n!$ is defined for integer values only.

    $$\left\{\frac{n!}{n^n}\right\} = \frac{1\times2\times3\times4\times...\times n}{n\times n\times n\times n\times...\times n}$$

2. This expression is less than $\frac{1}{n}$ but is also greater than $\frac{-1}{n}$.

3. Because the two sequences $\left\{\frac{1}{n}\right\}$ and $\left\{\frac{-1}{n}\right\}$ each converge to 0, we can apply the Squeeze Theorem to conclude that $\left\{\frac{n!}{n^n}\right\}$ converges to 0 also.

CRITICAL POINT

There is a theorem that states $\lim\limits_{n\to\infty} |a_n| = 0$ then $\lim\limits_{n\to\infty} a_n = 0$. This theorem is especially useful when dealing with sequences in the form $\{(-1)^n a_n\}$.

Example 11: Determine if the sequence $\left\{\left(\frac{-1}{2}\right)^n\right\}$ converges.

Solution: Because $\left\{\left(\frac{1}{2}\right)^n\right\}$ converges, so does $\left\{\left(\frac{-1}{2}\right)^n\right\}$.

> **YOU'VE GOT PROBLEMS**
>
> Problem 2: Show that the sequence $\left\{\frac{(-1)^n n}{n+1}\right\}$ converges.

Increasing, Decreasing, and Monotonic Sequences

As much as the language is annoying, strictly or not, we must know how to handle consecutive terms being equal. A sequence that is strictly increasing or strictly decreasing is called *monotonic* (not monotonous):

- A sequence is said to be strictly increasing if $a_{n+1} > a_n$ for all positive values of n.
- A sequence is said to be increasing if $a_{n+1} \geq a_n$ for all positive values of n.
- A sequence is said to be strictly decreasing if $a_{n+1} < a_n$ for all positive values of n.
- A sequence is said to be decreasing if $a_{n+1} \leq a_n$ for all positive values of n.

> **DEFINITION**
>
> A sequence that is strictly increasing or strictly decreasing is said to be **monotonic**.

There are two ways to prove sequences are strictly (or not) increasing or decreasing:

- If the sequence is a subset of a function with a domain defined as a subset of real numbers, we can use the derivative to determine whether the function increases or decreases.
- If the sequence is not defined as a subset of real numbers, we will look at the ratio of $\frac{a_{n+1}}{a_n}$ to determine if the values for which the ratio is greater than or less than 1.

Example 12: Show that the sequence $\left\{\frac{2n}{n+3}\right\}$ is strictly increasing.

Solution: Use the derivative of the function $f(x) = \frac{2x}{x+3}$, $f'(x) = \frac{2(x+3)-2x(1)}{(x+3)^2} = \frac{6}{(x+3)^2}$. This derivative is positive all values of x (except -3, which is not part of the sequence's domain), so the sequence is strictly increasing.

Example 13: Show that the sequence $\left\{\frac{10^n}{(2n)!}\right\}$ is decreasing.

Solution:

1. The term $a_{n+1} = \frac{10^{n+1}}{(2(n+1))!}$, so the ratio $\frac{a_{n+1}}{a_n} = \frac{\frac{10^{n+1}}{(2n+2)!}}{\frac{10^n}{(2n)!}} = \frac{10^{n+1}}{(2n+2)!} \div \frac{10^n}{(2n)!} = \frac{10^{n+1}}{(2n+2)(2n+1)(2n)!} \times \frac{(2n)!}{10^n}$
    $= \frac{10}{(2n+2)(2n+1)}$.

2. The problem is to determine when $\frac{10}{(2n+2)(2n+1)} < 1$.

3. The denominator is a positive term, so we can multiply both sides of the inequality without changing the orientation of the inequality.

4. The inequality becomes $10 < 4n^2 + 6n + 2$ and then $0 < 4n^2 + 6n - 8$. When you solve this quadratic, you get $n = \frac{-6 \pm \sqrt{36 - 4(4)(-8)}}{8} = \frac{-6 \pm \sqrt{164}}{8}$.

5. The decimal solutions are -2.35 and 0.85.

6. The negative solution is not part of the sequence's domain.

7. The solution to the inequality is $n > 0.85$ so the sequence is decreasing for $n > 1$.

YOU'VE GOT PROBLEMS

Problem 3: Determine if the sequence $\{ne^{-2n}\}$ is increasing, strictly increasing, decreasing, or strictly decreasing.

The Least You Need to Know

- A convergent sequence is one for which $\lim\limits_{n \to \infty} a_n$ is finite.

- A divergent sequence is one for which $\lim\limits_{n \to \infty} a_n$ is either infinite or cannot be determined.

- The Squeeze Theorem can be used to prove the convergence of a sequence whose limit to infinity you might not be able to determine.

- You can determine if a sequence is strictly increasing or strictly decreasing or possibly just increasing or decreasing by comparing the term a_n to a_{n+1} or by applying the First Derivative Test to the function that defines the sequence.

CHAPTER 15

Infinite Series

Whereas a sequence is a listing of numbers that come from a defined pattern, a series is the sum of the terms of the sequence. In the next two chapters, we will examine infinite series. We'll lay down the basics for infinite series in this chapter and work further with some important series in Chapter 16.

In This Chapter

- Investigating infinite geometric series
- Studying tests of convergence: ratio, comparison, and integral
- Examining alternating series and absolute and conditional convergence
- Estimating sums of alternating series with a maximum error estimate

Infinite Geometric Series

A geometric sequence is defined by the rule $a_n = a_1 r^{n-1}$. For example, the sequence $a_n = 2(3)^{n-1}$ is 2, 6, 18, ... Do you see why 2 must be the first term? $a_1 = 2(3)^{1-1} = 2(3)^0 = 2$. The sum of the first six terms of this series is $S_6 = 2 + 6 + 18 + 54 + 162 + 486$. There are only a few terms in this series so we could add them up by hand. Imagine if we had to find the sum of the first 60 terms, you would want to use a formula. The process is called "eliminating the middle."

1. We'll multiply the equation for S_6 by the common ratio, 3, to get $3S_6 = 6 + 18 + 54 + 162 + 486 + 1458$.

2. We subtract this from the equation for S_6. $S_6 - 3S_6 = (2 + 6 + 18 + 54 + 162 + 486) - (6 + 18 + 54 + 162 + 486 + 1458)$ so that $-2S_6 = 2 - 1458$. (Do you see how all the middle terms are eliminated?)

3. Just so that we can get to a more general formula, we'll rewrite 1458 as $2(3)^6$. (Grab your calculator to verify that is correct.)

4. So $-2S_6 = 2 - 2(3)^6$ so that $S_6 = \frac{2-2(3)^6}{-2} = \frac{2\left(1-(3)^6\right)}{-2} = 728$.

In general, if the first term of the geometric sequence is a_1 and the common ratio is r, the sum of the first n terms of the series is $S_n = \frac{a_1(1-r^n)}{1-r}$. (We'll not compute the sum of the first 60 terms as it is in the neighborhood of 4×10^{27}.)

Example 1: Find the sum of the first 40 terms of the series $2400 + 1800 + 1350 + 1012.5 + \dots$.

Solution: The first is 2400 and the common ratio is $\frac{1800}{2400} = \frac{3}{4}$. Therefore, $S_{40} = \frac{2400\left(1-\left(\frac{3}{4}\right)^{40}\right)}{1-\frac{3}{4}} = 9599.903$.

You can see that as n gets larger, the value of a_n gets smaller. The question is "What happens to S_n?" The $\lim\limits_{n \to \infty} S_n = \lim\limits_{n \to \infty} \frac{a_1(1-r^n)}{1-r}$. As n gets larger, $r^n \to 0$ because $r < 1$.

> ✏️ **CRITICAL POINT**
>
> The sum of an infinite geometric series will converge to $S = \frac{a_1}{1-r}$, provided $|r| < 1$.

Example 2: Compute $16 + 8 + 4 + 2 + 1 + \ldots$.

Solution: The first term is 16 and the common ratio is $\frac{1}{2}$, so the sum is $S = \frac{16}{1-\frac{1}{2}} = 32$.

Example 3: Compute $16 - 8 + 4 - 2 + 1 \ldots$.

Solution: The first term is 16 and the common ratio is $\frac{-1}{2}$, so the sum is $S = \frac{16}{1-\left(\frac{-1}{2}\right)} = \frac{32}{3}$.

> **YOU'VE GOT PROBLEMS**
>
> Problem 1: Compute $36 - 24 + 16 - \frac{32}{3} + \ldots$.

Tests of Convergence

If all the terms in a series Σa_n are positive, the series is called a positive series. (As opposed to an alternating series that we will discuss later in this chapter.) There are a number of tests available to us to determine if a positive series is convergent.

We begin with a very simple, but important statement, called the Divergence Test.

> **CRITICAL POINT**
>
> If $\lim\limits_{n \to \infty} a_n$ does not equal 0, the series $\sum\limits_{n=1}^{\infty} a_n$ diverges.

Simply said, if the numbers being added to the series do not get infinitely small, the sum of the terms will continue to grow.

Example 4: Show that the series $\sum\limits_{n=1}^{\infty} n\sin\left(\frac{1}{n}\right)$ diverges.

Solution:

1. We'll go back to the familiar limit theorem, $\lim\limits_{x \to 0} \frac{\sin(x)}{x} = 1$.

2. If we replace x with $\frac{1}{n}$, $x \to 0$ becomes $\frac{1}{n} \to 0$ so that $n \to \infty$ making

   $$\lim_{n \to \infty} \frac{\sin\left(\frac{1}{n}\right)}{\frac{1}{n}} = \lim_{n \to \infty} n\sin\left(\frac{1}{n}\right) = 1.$$

3. Therefore, series $\sum\limits_{n=1}^{\infty} n\sin\left(\frac{1}{n}\right)$ diverges.

When the rule defining the series is a continuous function that can be integrated, you can apply improper integrals to test for the convergence or divergence of a series. This is called the *Integral Test*.

> **DEFINITION**
>
> If f(x) is a continuous, positive, decreasing function on $[1, \infty]$ with $f(n) = a_n$ for all positive integers n, then the series Σa_n converges if and only if $\int_1^\infty f(x)\, dx$ converges. This is called the **Integral Test.**

Example 5: Show that $\displaystyle\sum_{n=1}^{\infty} \frac{1}{n}$ diverges.

Solution: $\displaystyle\int_1^\infty \frac{1}{x}\, dx = \lim_{p\to\infty} \int_1^p \frac{1}{x}\, dx = \lim_{p\to\infty} \ln|x|\Big|_1^p = \lim_{p\to\infty}\left(\ln|p| - \ln|1|\right) = \infty$. Because the integral diverges, the series also diverges.

> **CRITICAL POINT**
>
> The series $\displaystyle\sum_{n=1}^{\infty} \frac{1}{n}$ is called the *harmonic series* and is related to musical harmonies. To give you a sense of the size of "large" in mathematics, it takes 12,368 terms before the sum of the terms in the harmonic series reaches 10 and it takes more than 1.509×10^{43} terms to reach 100. That is pretty incredible!

Example 6: Show that $\displaystyle\sum_{n=1}^{\infty} \frac{1}{n^2}$ converges.

Solution: $\displaystyle\int_1^\infty \frac{1}{x^2}\, dx = \lim_{p\to\infty} \int_1^p \frac{1}{x^2}\, dx = \lim_{p\to\infty} \frac{-1}{x}\Big|_1^p = \lim_{p\to\infty}\left(\frac{-1}{p} - (-1)\right) = 1$. Because the integral converges, the series also converges.

> **CRITICAL POINT**
>
> An important consequence of the Integral Test is referred to as the *p*-series. (We've seen this before when we discussed improper integrals.) Series of the form $\displaystyle\sum_{n=1}^{\infty} \frac{1}{n^p}$ will converge when $p > 1$ and will diverge when $p \le 1$.

YOU'VE GOT PROBLEMS

Problem 2: Show that the series $\sum\limits_{n=1}^{\infty} \frac{n}{e^n}$ converges.

The next test for us to look at is the *Comparison Test*. This is a good test to use when the unknown series looks like a known convergent or divergent series.

DEFINITION

If two positive series, $\sum\limits_{n=1}^{\infty} a_n$ and $\sum\limits_{n=1}^{\infty} b_n$, with the property that $a_k > b_k$ for $k \geq m$ then we can conclude two things: If the series $\sum\limits_{n=1}^{\infty} a_n$ converges, then so does the series $\sum\limits_{n=1}^{\infty} b_n$. And if the series $\sum\limits_{n=1}^{\infty} b_n$ diverges, then so does the series $\sum\limits_{n=1}^{\infty} a_n$. This is called the **Comparison Test.** In English, there comes a point where the terms in series $\sum\limits_{n=1}^{\infty} a_n$ are always greater than the terms in series $\sum\limits_{n=1}^{\infty} b_n$.

Example 7: Show that the series $\sum\limits_{n=1}^{\infty} \frac{1}{n^2+5}$ converges.

Solution:

1. This series looks strikingly similar to the series $\sum\limits_{n=1}^{\infty} \frac{1}{n^2}$, which is a known convergent series.

2. We now have to show that there comes a point in which $\frac{1}{n^2} > \frac{1}{n^2+5}$.

3. Both denominators are positive, so we can multiply the inequality by the common denominator without affecting the orientation of the inequality: $n^2 + 5 > n^2$.

4. This becomes $5 > 0$, which is always true.

5. Because the unknown series is less than the known series for all values of n, the unknown series, $\sum\limits_{n=1}^{\infty} \frac{1}{n^2+5}$ converges.

Example 8: Show that the series $\sum\limits_{n=1}^{\infty} \frac{1}{n+1}$ diverges.

Solution:

1. It seems reasonable to compare this series to the divergent harmonic series. When is $\frac{1}{n+1} > \frac{1}{n}$? The answer is never.

2. For positive values of n, n is never greater than $n + 1$.

3. So what do we do? We come up with a different divergent series. Rather than use $\sum\limits_{n=1}^{\infty} \frac{1}{n}$, use the series $\sum\limits_{n=1}^{\infty} \frac{1}{2n}$. When is $\frac{1}{n+1} > \frac{1}{2n}$? Always!

4. Multiply by the common denominator to get $2n > n + 1$ or that $n > 1$.

5. The unknown series is greater than the known divergent series so the series $\sum\limits_{n=1}^{\infty} \frac{1}{n+1}$ diverges.

(Did you notice that applying the Integral Test shows that the series $\sum\limits_{n=1}^{\infty} \frac{1}{n+1}$ diverges?)

Example 9: Determine whether the series $\sum\limits_{n=1}^{\infty} \frac{1}{n^2 + 5n - 1250}$ converges or diverges.

Solution:

1. Once again, the series $\sum\limits_{n=1}^{\infty} \frac{1}{n^2}$ will make for a good comparison.

2. When is $\frac{1}{n^2} > \frac{1}{n^2 + 5n - 1250}$?

3. Solve: $n^2 + 5n - 1250 > n^2$ becomes $5n - 1250 > 0$ or $n > 250$.

4. The inequality takes 250 terms before it holds true, but eventually it is true that $\frac{1}{n^2} > \frac{1}{n^2 + 5n - 1250}$.

5. The series $\sum\limits_{n=1}^{\infty} \frac{1}{n^2 + 5n - 1250}$ converges.

YOU'VE GOT PROBLEMS

Problem 3: Use the Comparison Test to determine if the series $\sum \frac{n+3}{n(n+1)(n+2)}$ converges or diverges.

The *Limit Comparison Test* also takes advantage of known convergent and divergent series in determining the nature of the convergence of an unknown series.

DEFINITION

Given two positive series $\sum_{n=1}^{\infty} a_n$ and $\sum_{n=1}^{\infty} b_n$ let $\lim_{n \to \infty} \frac{a_n}{b_n} = L$. If L is positive and finite, then both $\sum_{n=1}^{\infty} \frac{1}{n+\sqrt{n}}$ and $\sum_{n=1}^{\infty} b_n$ converge or diverge. This is the

Limit Comparison Test.

This test has the advantage that is saves us from doing some ugly algebra.

Example 10: Show that $\sum_{n=1}^{\infty} \frac{1}{n+\sqrt{n}}$ diverges.

Solution: Use the Limit Comparison Test with the divergent series $\sum_{n=1}^{\infty} \frac{1}{n}$.

$\lim_{n \to \infty} \frac{\frac{1}{n}}{\frac{1}{n+\sqrt{n}}} = \lim_{n \to \infty} \frac{n+\sqrt{n}}{n} = \lim_{n \to \infty} 1 + \frac{1}{\sqrt{n}} = 1$. Therefore, both series diverge.

Example 11: Show that the series $\sum_{k=1}^{\infty} \frac{1}{k^5 + 10k}$ converges. (Sorry, I got tired of using n as the variable.)

Solution: Compare this series to $\sum_{k=1}^{\infty} \frac{1}{k^5}$. $\lim_{k \to \infty} \frac{\frac{1}{k^5}}{\frac{1}{k^5+10k}} = \lim_{k \to \infty} \frac{k^5+10k}{k^5} = \lim_{k \to \infty} 1 + \frac{10}{k^4} = 1$. Therefore, both series converge.

YOU'VE GOT PROBLEMS

Problem 4: Determine whether the series $\sum_{k=1}^{\infty} \frac{8k^2}{\sqrt[3]{4k^7 + 9k^3}}$ is convergent or divergent.

Alternating Series

Now that we've developed ground rules for how to deal with series containing only positive terms, we take on the issue of series with alternating signs. This will be the basis for the work we do with power series in the next chapter.

> ✏️ **CRITICAL POINT**
>
> Let $a_n = (-1)^n b_n$ with $b_n > 0$ for all n. If (1) b_n is decreasing for all n, and (2) $\lim_{n \to \infty} b_n = 0$, then $\sum_{n=1}^{\infty} a_n$ is convergent.

This statement is called the Alternating Series Test.

Example 12: Show that the alternating harmonic series $\sum_{n=1}^{\infty} \frac{(-1)^n}{n}$ converges.

Solution: The series meets the condition that $\frac{1}{n}$ decreases (use the derivative if you want; it's always negative) and $\lim_{n \to \infty} \frac{1}{n} = 0$. Therefore, $\sum_{n=1}^{\infty} \frac{(-1)^n}{n}$ converges.

Example 13: Determine if the series $\sum_{n=1}^{\infty} \frac{(-1)^n n^3}{n^3 + 9}$ is convergent or divergent.
Solution:

1. We know $\lim_{n \to \infty} \frac{n^3}{n^3 + 9} = 1$, so the conditions of the Alternating Series Test are not met.

2. This does not tell us that the series is divergent, but it does tell us we need to try something else. At this point, the only something else we have is the Divergence Test. If $\lim_{n \to \infty} \frac{(-1)^n n^3}{n^3 + 9}$ does not equal 0, then the series diverges.

3. As you can see, $\lim_{n \to \infty} \frac{(-1)^n n^3}{n^3 + 9}$ will oscillate between −1 and 1 for very large values of n.

4. The limit fails to exist, so the limit does not equal 0.

5. We can use the Divergence Test to conclude that $\sum_{n=1}^{\infty} \frac{(-1)^n n^3}{n^3 + 9}$ is divergent.

> ➗✖️ **YOU'VE GOT PROBLEMS**
>
> Problem 5: Determine if the series $\sum \frac{\sin\left(\frac{(2n-1)\pi}{2}\right)}{\sqrt{n}}$ is convergent?

So we now know that the harmonic series is divergent but the alternating harmonic series is convergent. This leads us to consider the other series that are related by their absolute values.

> **CRITICAL POINT**
>
> Given the series $\sum_{n=1}^{\infty} a_n$, the series is said to be *absolutely convergent* if the series $\sum_{n=1}^{\infty} |a_n|$ converges. The series is said to be *conditionally convergent* if $\sum_{n=1}^{\infty} a_n$ converges but $\sum_{n=1}^{\infty} |a_n|$ diverges. If a series is absolutely convergent, it is convergent.

Therefore, the alternating harmonic series is conditionally convergent.

Example 14: Is the series $\sum_{n=1}^{\infty} \frac{(-1)^n}{n^4}$ absolutely convergent, conditionally convergent, or neither?

Solution: The series $\left| \sum_{n=1}^{\infty} \frac{(-1)^n}{n^4} \right| = \sum_{n=1}^{\infty} \frac{1}{n^4}$ is convergent based on the *p*-series. Therefore, $\sum_{n=1}^{\infty} \frac{(-1)^n}{n^4}$ is absolutely convergent.

Example 15: Is the series $\sum_{n=1}^{\infty} \frac{\cos(n)}{n}$ absolutely convergent, conditionally convergent, or neither?

Solution:

1. The series itself does not represent an alternating series.

2. However, if we consider that $-1 \le \cos(n) \le 1$, then $\left| \frac{\cos(n)}{n} \right| \le \left| \frac{1}{n} \right|$.

3. Use the Comparison Test to note that $\sum_{n=1}^{\infty} \frac{(-1)^n}{n}$ converges conditionally then so does $\sum_{n=1}^{\infty} \frac{\cos(n)}{n}$.

> **CRITICAL POINT**
>
> Given a series $\sum_{n=1}^{\infty} a_n$, let $\lim_{n \to \infty} \left| \frac{a_{n+1}}{a_n} \right| = r$. If $r < 1$, then $\sum_{n=1}^{\infty} a_n$ converges. If $r > 1$, then $\sum_{n=1}^{\infty} a_n$ diverges. If $r = 1$, then no conclusion about $\sum_{n=1}^{\infty} a_n$ can be drawn from this test. This is called the Ratio Test.

The Ratio Test is a major player in the study of alternating series. It is especially useful for series involving nth powers and factorials.

Example 16: Determine if the series $\sum_{n=1}^{\infty} \frac{(-1)^n 2^{2n}}{(2n)!}$ is absolutely convergent, conditionally convergent, or neither.

Solution: Using the Ratio Test, $\lim_{n \to \infty} \left| \frac{\frac{2^{2(n+1)}}{(2(n+1))!}}{\frac{2^{2n}}{(2n)!}} \right| = \lim_{n \to \infty} \left| \frac{2^{2n+2}}{(2n+2)!} \times \frac{(2n)!}{2^{2n}} \right| = \lim_{n \to \infty} \left| \frac{4}{(2n+2)(2n+1)} \right| = 0$. By the terms of the Ratio Test, $\sum_{n=1}^{\infty} \frac{(-1)^n 2^{2n}}{(2n)!}$ is absolutely convergent.

Example 17: Determine if the series $\sum_{n=1}^{\infty} \frac{n^n}{n!}$ is absolutely convergent, conditionally convergent, or neither.

Solution: $\lim_{n \to \infty} \left| \frac{a_{n+1}}{a_n} \right| = \lim_{n \to \infty} \left| \frac{(n+1)^{n+1}}{(n+1)!} \times \frac{n!}{n^n} \right| = \lim_{n \to \infty} \left| \frac{(n+1)(n+1)^n}{(n+1)n!} \times \frac{n!}{n^n} \right| = \lim_{n \to \infty} \left| \frac{(n+1)^n}{n^n} \right| = \lim_{n \to \infty} \left| \left(\frac{n+1}{n} \right)^n \right| = \lim_{n \to \infty} \left| \left(1 + \frac{1}{n} \right)^n \right| = e$. With $e > 0$, the series diverges.

YOU'VE GOT PROBLEMS

Problem 6: Determine if the series $\sum_{n=1}^{\infty} \frac{n 4^n}{n!}$ is absolutely convergent, conditionally convergent, or neither.

The Root Test is the last of the tests for convergence/divergence that we need to consider.

CRITICAL POINT

Given a series $\sum_{n=1}^{\infty} a_n$, let $\lim_{n \to \infty} \sqrt[n]{|a_n|} = r$. If $r < 1$, then $\sum_{n=1}^{\infty} a_n$ is absolutely convergent. If $r > 1$, then $\sum_{n=1}^{\infty} a_n$ divergent. If $r = 1$, then no conclusion about $\sum_{n=1}^{\infty} a_n$ can be drawn from this test. This is called the Root Test.

A useful fact to have when using the Root Test is $\lim_{n \to \infty} \sqrt[n]{n} = 1$.

Example 18: Determine if the series $\sum_{n=1}^{\infty}\left(\frac{4-n^5}{9n^5+7}\right)^n$ is absolutely convergent, conditionally convergent, or neither.

Solution: Applying the Root Test, $\lim_{n \to \infty} \sqrt[n]{\left|\left(\frac{4-n^5}{9n^5+7}\right)^n\right|} = \lim_{n \to \infty}\left|\frac{4-n^5}{9n^5+7}\right| = \lim_{n \to \infty}\left|\frac{\frac{4}{n^5}-1}{9+\frac{7}{n^5}}\right| = \left|\frac{-1}{9}\right| = \frac{1}{9}$. The series is absolutely convergent.

Estimating the Sum of Alternating Series

Let's reinforce two items about alternating series that will help with the work we are going to do on estimating sums.

- If we define the alternating series as $\sum_{n=1}^{\infty}(-1)^n a_n$, the terms a_n are positive and they are decreasing.

- It is the factor $(-1)^n$ that causes the signs to alternate.

The sum of the first n terms of the series, called the partial sum s_n, estimates the true sum. However, there is a difference between partial sum and the true sum. The maximum value of this difference is the next term in the series, a_{n+1}.

> **CRITICAL POINT**
>
> The difference between the sum of an alternating series, S, and a partial sum, s_n, is at most a_{n+1}. That is, $\left|S - s_n\right| < a_{n+1}$.

Example 19: Estimate the maximum error when $\sum_{n=1}^{\infty}\frac{(-1)^n}{n^2}$ is estimated by s_{15}.

Solution: $\left|S - s_{15}\right| < a_{16}$ so the maximum error is $\frac{1}{16^2} = \frac{1}{256}$.

Example 20: How many terms are needed to estimate the sum $\sum_{k=1}^{\infty}\frac{(-1)^k}{k!}$ so that the maximum error is less than 0.0001?

Solution:

1. The maximum error is $\left|S - s_k\right| < a_{k+1} < 0.0001$.

2. $\frac{1}{(k+1)!} < \frac{1}{10000}$ becomes $10000 < (k+1)!$.

3. Use your calculator to determine that $k + 1 = 8$ so $k = 7$.

4. Use seven terms to insure that the maximum error is less than 0.0001.

5. The sum $\sum_{k=1}^{\infty} \frac{(-1)^k}{k!} = \frac{1}{e} - 1$, which is approximately -0.632121. $\sum_{k=1}^{7} \frac{(-1)^k}{k!}$ is -0.632143.

6. The first four decimal places are exactly the same.

YOU'VE GOT PROBLEMS

Problem 7: Determine the number of terms needed to estimate the sum

$\sum_{n=1}^{\infty} \frac{n^2}{2^n}$ so that the maximum error is less than 0.00001.

The Least You Need to Know

- Look to see if $\lim_{n \to \infty} a_n$ is equal to 0. If it's not, the series diverges by the Divergence Test.

- A geometric series converges if $|r| < 1$; the series diverges if $|r| > 1$.

- Use the Integral Test when the series has the form of a function that can be integrated.

- Use the Comparison Test when the series looks like a p-series.

- When the series involves the ratio of polynomials, use the Comparison Test or Limit Comparison Test after checking that all the terms of the series are positive.

- Use the Ratio Test when the series involve factorials or constants raised to powers.

- Use the Alternating Series Test when the series have the form $(-1)^n a_n$.

Power Series

We use power series to create polynomial approximations for nonpolynomial functions. In today's world, in which we can compute almost anything on our calculators, this might seem like another out-of-date topic. You will see the complete opposite is actually true—it's because of the application of power series that programmers can use these polynomials to compute values of functions such as sine, cosine, and the natural logarithm.

In This Chapter

- Working with the power series

- Interval and radius of convergence

- Computing the MacLaurin Series

- Working with the Taylor Series

- Estimating errors for the Taylor and MacLaurin Series

Power Series

Power series take on the form $\sum_{n=0}^{\infty} a_n(x-c)^n$ where c is some constant. This series will always converge when $x = c$ and the result would be a_0. What about the other values of x? It could be the case that the series converges for all values of x. It might be the case that the series converges for some values of x with c in the middle of the interval. (We just covered the case that it converges for no other value of x in Chapter 15.) There are two concepts we need to clarify in this chapter—the *radius of convergence* and the *interval of convergence*.

> **DEFINITION**
>
> The **radius of convergence** identifies how far from $x = c$ one can go to find a value of x for which the series converges. The **interval of convergence** names the values of x for which the series converges.

Example 1: Determine the radius of convergence and interval of convergence for the power series $\sum_{n=1}^{\infty} \frac{x^n}{n4^n}$.

Solution:

1. The value of c for this problem is 0 (when you set $x - c = x$ you get $c = 0$).

2. We'll use the Ratio Test to determine when the series converges. Remember, we need $\lim_{n\to\infty} \left| \frac{a_{n+1}}{a_n} \right| < 1$.

3. $\lim_{n\to\infty} \left| \frac{x^{n+1}}{(n+1)4^{n+1}} \times \frac{n4^n}{x^n} \right| = \lim_{n\to\infty} \left| \frac{nx}{(n+1)4} \right| = \left| \frac{x}{4} \right| < 1$, which implies that $|x| < 4$ or that $-4 < x < 4$.

4. The radius of convergence is 4. To determine the interval of convergence, we need to take a look at what happens when the limiting value from the Ratio Test is 1.

5. We need to test the series for $x = -4$ and $x = 4$.

6. When $x = 4$, the series becomes $\sum_{n=1}^{\infty} \frac{4^n}{n4^n} = \sum_{n=1}^{\infty} \frac{1}{n}$. This is the harmonic series and we know that it is divergent.

7. When $x = -4$, the series becomes $\sum_{n=1}^{\infty} \frac{(-4)^n}{n4^n} = \sum_{n=1}^{\infty} \frac{(-1)^n}{n}$. This is the alternating harmonic series and we know that it is convergent.

8. Therefore, the interval of convergence is $[-4,4)$.

BE AWARE

When testing for the interval of convergence of a power series, you must always test the endpoints separately to see if the series converges at these points.

Example 2: Determine the radius of convergence and interval of convergence for the power series $\sum\limits_{n=0}^{\infty} \frac{x^n}{n!}$.

Solution: The Ratio Test gives $\lim\limits_{n\to\infty} \left| \frac{x^{n+1}}{(n+1)!} \times \frac{n!}{x^n} \right| = \lim\limits_{n\to\infty} \left| \frac{x}{n+1} \right| = 0$. The radius of convergence is ∞.

Therefore, the interval of convergence is $(-\infty,\infty)$.

Example 3: A function f(x) is defined as $f(x) = \frac{1}{4} + \frac{2}{4^2}x + \frac{3}{4^3}x^2 + \frac{4}{4^4}x^3 + \dots + \frac{n+1}{4^{n+1}}x^n$.

1. Find the interval of convergence for f(x).

2. Write the first three terms and the general term for $\int_0^1 f(x)\, dx$.

3. Find the sum of the series found in 2.

Solution:

1. Use the Ratio Test to get $\lim\limits_{n\to\infty} \left| \frac{(n+2)x^{n+1}}{4^{n+2}} \times \frac{4^{n+1}}{(n+1)x^n} \right| = \lim\limits_{n\to\infty} \left| \frac{(n+2)x}{4(n+1)} \right| = \left| \frac{x}{4} \right|$. If $\left| \frac{x}{4} \right| < 1$, then

    $-4 < x < 4$. Test the endpoints. When $x = 4$, the series becomes $\sum \frac{(n+1)4^n}{4^{n+1}} = \sum \frac{(n+1)}{4}$,

    which diverges by the Divergence Test. When $x = -4$, the series becomes

    $\sum \frac{(n+1)(-4)^n}{4^{n+1}} = \sum \frac{(-1)^n(n+1)}{4}$, which diverges by the Divergence Test. Therefore, the

    interval of convergence is $(-4,4)$.

2. You integrate the power series term by term as you would any polynomial:

    $\int_0^1 f(x)\, dx = \int_0^1 \frac{1}{4} + \frac{2}{4^2}x + \frac{3}{4^3}x^2 + \frac{4}{4^4}x^3 + \dots + \frac{n+1}{4^{n+1}}x^n \, dx =$

    $\frac{1}{4}x + \frac{1}{4^2}x^2 + \frac{1}{4^3}x^3 + \frac{1}{4^4}x^4 + \dots + \frac{1}{4^{n+1}}x^{n+1} \Big|_0^1 = \frac{1}{4} + \frac{1}{4^2} + \frac{1}{4^3} + \dots + \frac{1}{4^{n+1}}$

3. This is an infinite geometric series, and the sum is $\frac{\frac{1}{4}}{1-\frac{1}{4}} = \frac{1}{3}$.

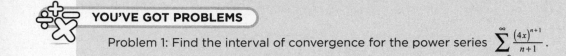

YOU'VE GOT PROBLEMS

Problem 1: Find the interval of convergence for the power series $\sum\limits_{n=0}^{\infty} \dfrac{(4x)^{n+1}}{n+1}$.

MacLaurin Series

Let's assume that the function f(x) has a power series associated with it and that power series is

$$f(x) = a_0 + a_1 x + a_2 x^2 + a_3 x^3 + \ldots + a_n x^n + \ldots = \sum_{n=0}^{\infty} a_n x^n.$$ The question then becomes "How do we

determine the coefficients of the terms in the power series?"

1. We know that f(0) = a_0 because all the other terms in the series will have a factor of 0.

2. If we take the derivative of the power series (which is done term by term), we get
   $$f'(x) = a_1 + 2a_2 x + 3a_3 x^2 + \ldots + n a_n x^{n-1} + \ldots = \sum_{n=0}^{\infty} n a_n x^{n-1}$$ and we now know that
   f'(0) = a_1.

3. Take the second derivative:
   $$f''(x) = 2a_2 + 6a_3 x + \ldots + n(n-1)a_n x^{n-2} + \ldots = \sum_{n=0}^{\infty} n(n-1)a_n x^{n-2}$$ and find that f''(0) =
   $2a_2$ so that $a_2 = \frac{f''(0)}{2}$.

4. If we continue in this manner, we find that the *n*th derivative evaluated at 0 gives the value of the *n*th coefficient.

5. That is, $a_n = \frac{f^{(n)}(0)}{n!}$.

6. Technically speaking, there is no zero derivative. The term $f^{(0)}(0)$ is the value of the function at x = 0.

DEFINITION

The **MacLaurin Series** for f(x) is a power series about x = 0 so that
$$f(x) = f(0) + f'(0)x + \frac{f''(0)}{2!}x^2 + \frac{f'''(0)}{3!}x^3 + \ldots + \frac{f^{(n)}(0)}{n!}x^n + \ldots = \sum_{n=0}^{\infty} \frac{f^{(n)}(0)}{n!}x^n .$$

Example 4: Determine the MacLaurin Series for $f(x) = \sin(x)$.

Solution:

1. We have $f(0) = 0$. $f'(x) = \cos(x)$, so $f'(0) = 1$. $f''(x) = -\sin(x)$ and $f''(0) = 0$.

2. The third derivative of $\sin(x)$ is $-\cos(x)$ and $f'''(0) = -1$.

3. This pattern is going to repeat itself.

4. The values of the odd derivatives will alternate between -1 and 1, while the values of the function and even derivatives will all be 0.

5. Therefore, the MacLaurin Series for $f(x) = \sin(x) = x - \frac{x^3}{3!} + \frac{x^5}{5!} - \frac{x^7}{7!} + \ldots + \frac{(-1)^n x^{2n+1}}{(2n+1)!}$.

How good is this fit? The graph of $f(x) = \sin(x)$ is sketched on the interval $[-2\pi, 2\pi]$ along with the MacLaurin Series for various number of terms, as shown in the following figures.

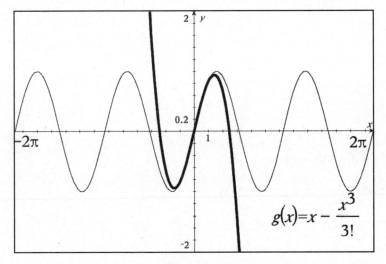

Figure 16.1
$f(x) = \sin(x)$ *and* $f(x) = x - \frac{x^3}{3!}$.

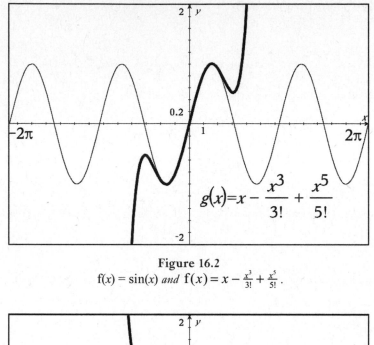

Figure 16.2

$f(x) = \sin(x)$ *and* $f(x) = x - \frac{x^3}{3!} + \frac{x^5}{5!}$.

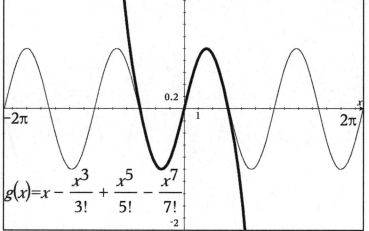

Figure 16.3

$f(x) = \sin(x)$ *and* $f(x) = x - \frac{x^3}{3!} + \frac{x^5}{5!} - \frac{x^7}{7!}$.

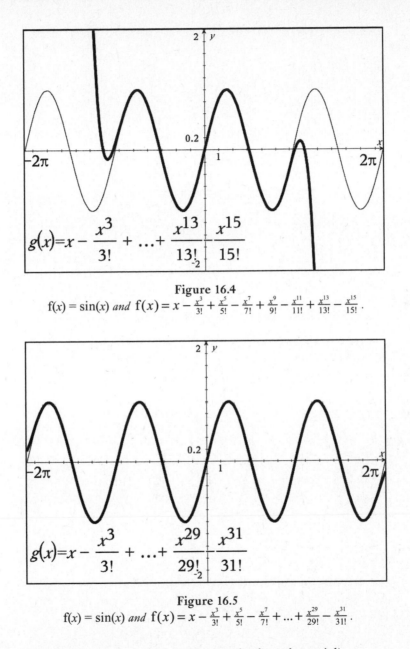

$$g(x)=x-\frac{x^3}{3!}+...+\frac{x^{13}}{13!}-\frac{x^{15}}{15!}$$

Figure 16.4

$f(x) = \sin(x)$ *and* $f(x) = x - \frac{x^3}{3!} + \frac{x^5}{5!} - \frac{x^7}{7!} + \frac{x^9}{9!} - \frac{x^{11}}{11!} + \frac{x^{13}}{13!} - \frac{x^{15}}{15!}$.

$$g(x)=x-\frac{x^3}{3!}+...+\frac{x^{29}}{29!}-\frac{x^{31}}{31!}$$

Figure 16.5

$f(x) = \sin(x)$ *and* $f(x) = x - \frac{x^3}{3!} + \frac{x^5}{5!} - \frac{x^7}{7!} + ... + \frac{x^{29}}{29!} - \frac{x^{31}}{31!}$.

That is pretty incredible accuracy for just 15 terms in the polynomial!

Example 5: Determine the interval of convergence for the MacLaurin Series for $f(x) = \sin(x)$.

Solution:

1. Use the Ratio Test to get $\lim\limits_{n \to \infty} \left| \dfrac{(-1)^{n+1} x^{2(n+1)+1}}{(2(n+1)+1)!} \times \dfrac{(2n+1)!}{(-1)^n x^{2n+1}} \right|$. (Actually, the terms with the -1 are not necessary because of the absolute value, but I didn't want to confuse you with that on the first example.)

2. The limit becomes $\lim\limits_{n \to \infty} \left| \dfrac{x^{2n+3}}{(2n+3)!} \times \dfrac{(2n+1)!}{x^{2n+1}} \right| = \lim\limits_{n \to \infty} \left| \dfrac{x^2}{(2n+3)(2n+2)} \right| = 0$.

3. The series converges for all values of x so the interval of convergence is $(-\infty, \infty)$.

YOU'VE GOT PROBLEMS

Problem 2: Show that the MacLaurin Series for $f(x) = \cos(x)$ is

$1 - \dfrac{x^2}{2} + \dfrac{x^4}{4!} - \dfrac{x^6}{6!} + \ldots + \dfrac{(-1)^n x^{2n}}{(2n)!}$.

Example 6: Determine the MacLaurin Series for $f(x) = e^x$ and find the interval of convergence.

Solution:

1. All the derivatives of e^x are e^x, so the values of all the derivatives at $x = 0$ are 1.

2. Therefore, $e^x = 1 + x + \dfrac{x^2}{2} + \dfrac{x^3}{3!} + \ldots + \dfrac{x^n}{n!}$.

3. Using the Ratio Test, $\lim\limits_{n \to \infty} \left| \dfrac{x^{n+1}}{(n+1)!} \times \dfrac{n!}{x^n} \right| = \lim\limits_{n \to \infty} \left| \dfrac{x}{n+1} \right| = 0$ so the interval of convergence is $(-\infty, \infty)$.

CRITICAL POINT

Your calculator is programmed to use power series to evaluate the trigonometric and logarithmic functions. In terms of computing, many algorithms help you evaluate polynomials in very little time.

There is a relationship among the MacLaurin Series for sin(x), cos(x), and e^x that is extremely powerful. The sum of cos(x) and $i$sin(x) (where i is $\sqrt{-1}$) is e^{ix}.

$$\cos(x) + i\sin(x) = 1 + ix - \frac{x^2}{2} - \frac{x^3}{3!}i + \frac{x^4}{4!} + \frac{x^5}{5!}i - \frac{x^6}{6!} - \frac{x^7}{7!}i + \frac{x^8}{8!} + \frac{x^9}{9!}i + \ldots$$

$$e^{ix} = 1 + ix + \frac{(ix)^2}{2} + \frac{(ix)^3}{3!} + \frac{(ix)^4}{4!} + \frac{(ix)^5}{5!} + \frac{(ix)^6}{6!} + \frac{(ix)^7}{7!} + \frac{(ix)^8}{8!} + \frac{(ix)^9}{9!} + \ldots =$$

$$1 + ix - \frac{x^2}{2} - \frac{x^3}{3!}i + \frac{x^4}{4!} + \frac{x^5}{5!}i - \frac{x^6}{6!} - \frac{x^7}{7!}i + \frac{x^8}{8!} + \frac{x^9}{9!}i + \ldots$$

Evaluating this relationship with $x = \pi$, $e^{i\pi} = \cos(\pi) + i\sin(\pi) = -1$ or $e^{i\pi} + 1 = 0$. The equation contains the five numbers upon which all of mathematics is based.

One other relationship that is important is $e^{\frac{i\pi}{2}} = \cos\left(\frac{\pi}{2}\right) + i\sin\left(\frac{\pi}{2}\right) = i$. Raise both sides of the equation to the i power to get $\left(e^{i^2\frac{\pi}{2}}\right) = i^i$ or $e^{\frac{-\pi}{2}} = i^i$. You can use your calculator to get the decimal approximation for this number. Why is it important? The development of mathematics is the development of mankind. Look at the names we have given the sets of numbers over the years.

- **Natural (counting) numbers**   We started with the natural (counting) numbers (1, 2, 3, …) as we counted our fingers, cattle, and so on.

- **Whole numbers**   Reality set in and we realize that we could have nothing, so we have the whole numbers (0, 1, 2, 3, …).

- **Rationals and irrationals**   We get the rationals (ratios, not statements about one's sanity) and then, of course, the irrationals.

- **Real numbers**   Together, the rationals and irrationals form the real numbers—all the numbers we can see in our real world.

- **Imaginary numbers**   And then along comes $\sqrt{-1}$, not something you find in the real world. Of course, we called these number imaginary numbers.

- **Complex numbers**   The combination of real and imaginary numbers.

Given that none of the numbers are real, can you show someone a 1? Not a finger or a pencil but a 1. In the same way that all of language is an abstract, so are the numbers that we used. The moral of this dialogue, all of the numbers are related to each other.

CRITICAL POINT

These are important MacLaurin Series that you should know and the interval of convergence for each.

$$\sin(x) = x - \frac{x^3}{3!} + \frac{x^5}{5!} - \frac{x^7}{7!} + \ldots + \frac{(-1)^n x^{2n+1}}{(2n+1)!} \text{ on } (-\infty,\infty)$$

$$\cos(x) = 1 - \frac{x^2}{2} + \frac{x^4}{4!} - \frac{x^6}{6!} + \ldots + \frac{(-1)^n x^{2n}}{(2n)!} \text{ on } (-\infty,\infty)$$

$$e^x = 1 + x + \frac{x^2}{2} + \frac{x^3}{3!} + \ldots + \frac{x^n}{n!} \text{ on } (-\infty,\infty)$$

$$\tan^{-1}(x) = x - \frac{x^3}{3} + \frac{x^5}{5} - \frac{x^7}{7} + \ldots + \frac{(-1)^n x^{2n+1}}{2n+1} \text{ on } [-1,1]$$

$$\frac{1}{1-x} = 1 + x + x^2 + x^3 + x^4 + \ldots + x^n \text{ on } (-1,1)$$

$$\ln(x+1) = x - \frac{x^2}{2} + \frac{x^3}{3} - \frac{x^4}{4} + \ldots + \frac{(-1)^{n+1} x^n}{n} \text{ on } (-1,1]$$

Example 7: Determine the MacLaurin Series for $f(x) = x^3 e^x$.

Solution: The MacLaurin Series for e^x is $\displaystyle\sum_{n=0}^{\infty} \frac{x^n}{n!}$ so the series for $f(x) = x^3 e^x$ is

$$\sum_{n=0}^{\infty} \frac{x^n}{n!} x^3 = \sum_{n=0}^{\infty} \frac{x^{n+3}}{n!} = x^3 + x^4 + \frac{x^5}{2} + \frac{x^6}{3!} + \frac{x^7}{4!} + \ldots + \frac{x^{n+3}}{n!}.$$

YOU'VE GOT PROBLEMS

Problem 3: Use the MacLaurin Series for $\cos(x)$ to find $\displaystyle\int \cos(x)\, dx$.

Taylor Series

The MacLaurin Series requires that the series be built about $x = 0$. The Taylor Series is more general than that in that it allows the series to be built around any value $x = c$.

DEFINITION

The Taylor Series for $f(x)$ about $x = c$ is $\displaystyle\sum_{n=0}^{\infty} \frac{f^{(n)}(c)(x-c)^n}{n!}$.

Example 8: Find the first three terms and the general term of the Taylor Series for $f(x) = \ln(x)$ about $x = 2$.

Solution:

1. The series is $\displaystyle\sum_{n=0}^{\infty} \frac{f^{(n)}(2)(x-2)^n}{n!}$.

2. The first term is $f(2) = \ln(2)$.

3. Let's look at the derivatives for $\ln(x)$ evaluated at $x = 2$:

    $f'(x) = \frac{1}{x}$ gives $f'(2) = \frac{1}{2}$

    $f''(x) = \frac{-1}{x^2}$ gives $f''(2) = \frac{-1}{2^2}$

    $f'''(x) = \frac{2}{x^3}$ gives $f'''(2) = \frac{2}{2^3}$

    $f^{(4)}(x) = \frac{-3!}{x^4}$ gives $f^{(4)}(2) = \frac{-3!}{2^4}$

    $f^{(5)}(x) = \frac{4!}{x^5}$ gives $f^{(5)}(2) = \frac{4!}{2^5}$

4. It looks like the signs alternate with the even derivatives being negative and the odd derivatives being positive.

5. The denominators are 2 raised to the same power as the order of the derivative and the numerator is the factorial for one less than the order of the derivative.

6. This all works except for the case in which $n = 0$. So we "cheat" and separate the first term from the rest of the batch. $\ln(x) = \ln(2) + \displaystyle\sum_{n=1}^{\infty} \frac{(-1)^{n+1}(n-1)!}{2^n n!}(x-2)^n = \sum_{n=1}^{\infty} \frac{(-1)^{n+1}}{2^n n}(x-2)^n$

    $= \ln(2) + \frac{1}{2}(x-2) - \frac{1}{2^2 \cdot 2}(x-2)^2 + \dots + \frac{(-1)^{n+1}}{2^n \cdot n}(x-2)^n$.

YOU'VE GOT PROBLEMS

Problem 4: Find the first three terms and the general term of the Taylor Series for $f(x) = \ln(x)$ about $x = 1$.

Example 9: The Taylor Series for a certain function f converges to $f(x)$ for all x in the interval of convergence. The nth derivative at $x = 0$ is given by:

$$f^{(n)}(0) = \frac{(-1)^n(n+1)!}{4^n(n+2)^2} \quad \text{for all } n$$

Write a fourth degree Taylor polynomial, $T(x)$, for f about $x = 0$ given that $f(0) = 8$ and determine the radius of convergence for f about $x = 0$.

Solution:

1. We have $f(0) = 8$, $f'(0) = \frac{-1(2)!}{4(9)} = \frac{-1}{9}$, $f''(0) = \frac{1(3)!}{16(16)} = \frac{3}{128}$, $f'''(0) = \frac{-1(4)!}{64(25)} = \frac{-3}{200}$, and

    $f^{(4)}(0) = \frac{1(5)!}{256(36)} = \frac{5}{384}$.

2. The coefficients for the terms in the polynomial are $\frac{f^{(n)}(0)}{n!}$, so the fourth degree Taylor polynomial is $T(x) = 8 - \frac{1}{18}x + \frac{3}{256}x^2 - \frac{1}{400}x^3 + \frac{5}{9216}x^4$.

3. Use the Ratio Test to determine the radius of convergence. Remember to include the terms for x as part of this test.

4. $\lim\limits_{n \to \infty} \left| \frac{(n+2)x^{n+1}}{4^{n+1}(n+3)^2} \times \frac{4^n(n+2)^2}{(n+1)x^n} \right| = \lim\limits_{n \to \infty} \left| \frac{n+2}{n+1} \left(\frac{n+2}{n+3} \right)^2 \frac{x}{4} \right| = \left| \frac{x}{4} \right| < 1$, which becomes $|x| < 4$. The radius of convergence is 4.

BE AWARE

You do not need to check the endpoints of the interval because you were not asked to determine the interval of convergence.

Error Estimates for the MacLaurin and Taylor Series

We saw in Chapter 15 that the error estimate for the sum of an alternating series is $\left| S - s_n \right| < a_{n+1}$ where:

- S is the true sum.

- s_n is the sum of the first n terms.

- a_{n+1} is the next term in the series.

In the same way, the difference between the true value of f(x_0) and T(x_0), the Taylor approxima-

tion for f(x_0), is given by $\left| f\left(x_0\right) - T_n\left(x_0\right) \right| < T_{n+1}\left(x_0\right)$, where $T_{n+1}\left(x_0\right) = \frac{f^{(n+1)}\left(x_0 - c\right)^{n+1}}{(n+1)!}$. This is

called the *Lagrange Error Estimate for the Taylor and MacLaurin Series*.

DEFINITION

The **Lagrange Error Estimate for the Taylor and MacLaurin Series** is

$$\left| f\left(x_0\right) - T_n\left(x_0\right) \right| < T_{n+1}\left(x_0\right), \text{ where } T_{n+1}\left(x_0\right) = \frac{f^{(n+1)}\left(x_0 - c\right)^{n+1}}{(n+1)!} .$$

Example 10: How many terms are needed to approximate $\sin\left(\frac{\pi}{3}\right)$ so that the error is less than 0.000001?

Solution:

1. We know that for all derivatives of sin(x), $\left| f^{(n)}(x) \right| \le 1$.

2. Therefore, $\left| T_{n+1}\left(x_0\right) \right| \le \left| \frac{\left(x_0 - c\right)^{n+1}}{(n+1)!} \right|$.

3. The sine function is built about x = 0, so we know that c = 0. This gives

    $$\left| T_{n+1}\left(\frac{\pi}{3}\right) \right| \le \left| \frac{\left(\frac{\pi}{3}\right)^{n+1}}{(n+1)!} \right| .$$

4. Use the list and spreadsheet feature on your calculator to determine the number of terms

    needed before $\left| \frac{\left(\frac{\pi}{3}\right)^{n+1}}{(n+1)!} \right|$ is less than 0.000001.

5. The value of this expression after 9 terms is approximately 4×10^{-7}.

Example 11: The Taylor Series for $f(x) = e^{-x^2}$ is $T(x) = 1 - x^2 + \frac{x^4}{2} - \frac{x^6}{6} + \ldots + \frac{(-1)^n x^{2n}}{n!}$.

Use the first two terms of this series to compute $\int_0^{1/2} e^{-x^2}\, dx$ and explain why it differs from

$\int_0^{1/2} f(x)\, dx$ by less than $\frac{1}{200}$.

Solution: The approximation is $\int_0^{1/2} 1 - x^2\, dx = x - \frac{x^3}{3} \Big|_0^{1/2} = \frac{1}{2} - \frac{1}{24} = \frac{11}{24}$. The maximum error is less

than the next term of the series when evaluated at $\frac{1}{2}$, $\frac{\left(\frac{1}{2}\right)^5}{10} = \frac{1}{320} < \frac{1}{200}$.

YOU'VE GOT PROBLEMS

Problem 5: The MacLaurin Series for $f(x) = \frac{1}{1+x^4}$ is

$T(x) = 1 - x^4 + x^8 - x^{12} + \ldots + (-1)^n x^{4n}$ Use the first three terms of this series

to approximate $\int_0^{\frac{1}{2}} f(x)\,dx$ and explain why this estimate is within $\frac{1}{10000}$ of the exact value.

The Least You Need to Know

- Functions are approximated by power series on your calculator.
- To determine the radius of convergence and interval of convergence for power series, use the Ratio Test.
- To estimate the number of terms needed to approximate a functional value within a given tolerance, use the Lagrange Error Estimate.

CHAPTER 17

Calculus II Final Exam

Here is your opportunity to test yourself on the material we have covered in this book. If you find yourself struggling with a problem, skip it and finish the rest of the problems that go with your chapter. When you've finished those problems you can answer, do not look at the solutions yet. Go back to the chapter material and look at the material relating to the problem(s) that gave you trouble, and maybe you'll be better able to answer the problem(s). After you've done this, even if you could not complete all the problems, look at the solutions to see how you have done. Above all, have fun!

In This Chapter

- Practicing what you've learned
- Testing your knowledge of the formulas and their applications
- Checking your work with the end-of-chapter answers

Some thoughts for how you can work through the problems:

- When appropriate, draw a picture on paper or use your graphing calculator to get a visualization of the problem.

- Write appropriate formulas on your paper as you work through the problems.

- Try to work out the problems by hand first. Use technology to evaluate the derivative or integral only when you can't do the problem with pen and paper.

Chapter 1

1. Convert $120°$ to radians.

2. Given $\cos(B) = \frac{-3}{8}$ and $\sin(B) > 0$, find $\cos(2B)$.

3. Simplify $\ln\left(\dfrac{e^{5x}\sin^2(x)}{(x+3)^2\sqrt{x-1}}\right)$.

4. Rewrite the parametric equations $x = 5\tan(t)$ and $y = 3\sec(t)$ in terms of x and y only.

5. Find the rectangular coordinates for the point with polar coordinates $\left(12, \frac{-3\pi}{4}\right)$.

6. For what values of θ in the interval $[0, 2\pi]$, does the graph of $r = 2 - 4\cos(\theta)$ pass through the origin of the polar coordinate system?

7. Find the sum of the infinite series $24 - 12 + 6 - 3 + 1.\,5\, \ldots$.

8. Rewrite $\dfrac{11x + 50}{x^2 + 9x + 20}$ in terms of its component fractions.

Chapter 2

For Problems 9 through 12, evaluate the given limits.

9. $\lim\limits_{x \to 3} \dfrac{x^2 + 4x - 21}{x^2 - 9}$

10. $\lim\limits_{x \to 2} g(x)$ where $g(x) = \begin{cases} 2x - 1 & x < 2 \\ \dfrac{1}{2}x^2 + 1 & x > 2 \end{cases}$

11. $\lim\limits_{h \to 0} \dfrac{\cos(x + h) - \cos(x)}{h}$

12. $\lim\limits_{x \to \infty} \dfrac{4x^3 - 5x^2 + 12x - 1000}{8100 + 9x - 15x^2 - 12x^3}$

For Problems 13 through 16, find the derivative for each of the functions.

13. $f(x) = x^2\ln(\cos(x))$

14. $g(x) = \dfrac{\tan\left(x^2\right)}{\sin(2x)}$

15. $k(x) = \sin^{-1}\left(e^{\cos(x)}\right)$

16. $p(x) = \sqrt[3]{x^3 + 5} + \dfrac{1}{\sqrt{6x - 2}}$

17. Find $\frac{dy}{dx}$ if $\tan(x^2 y) + xy^2 = 0$.

18. A rectangle is inscribed within the graph of $y = e^{-x^2}$. One side of the rectangle lies on the x-axis and two of the vertices lie on the graph. Determine the maximum possible area of the rectangle.

19. A rectangle with fixed area of 80 square units has a width that increases at the rate of 4 units per minute.

     (a) Determine the rate of change of the length at the instant the width is 8 units.

     (b) Determine the rate of change of the length of the diagonal at the instant the width is 8 units.

Chapter 3

20. Given $f'(x) = 9x^4 + \sqrt{2x - 3}$, find f(x).

21. Given $\frac{dy}{dx} = \sqrt[3]{6x + 5} + e^{4x} - 3$, find an expression for y.

22. Evaluate $\int 5\cos(x) + \frac{1}{1+x^2} - 1\,dx$.

23. Evaluate $\int \frac{1}{\sqrt{1-x^2}} + \frac{1}{\sqrt{1-x}}\,dx$.

24. Given $g'(x) = \frac{4}{x+9} - x + 3$ and $g(-8) = 4$, find the function g(x).

25. Evaluate $\int \frac{3+x}{1+x^2}\,dx$.

26. Evaluate $\int_0^\pi x + \cos(x)\,dx$.

27. If $g(x) = \int_0^x \sin^3(t)\cos^2(t)\,dt$, find g'(x).

28. A particle moves along a horizontal line with velocity $v(t) = 4\sqrt{t+3} + t$. What is the displacement of the object over the time interval $0 \le t \le 6$?

29. Evaluate $\int_4^6 \frac{\sin(\ln(x-3))}{x-3}\, dx$.

Chapter 4

30. Evaluate $\int_{\frac{-\pi}{3}}^0 \tan(x) - \sqrt{3}x\, dx$.

31. Use the midpoint method to estimate $\int_0^4 \sqrt{2x+1}\, dx$ with 10 subdivisions.

32. Use Simpson's Rule to estimate $\int_0^4 \sqrt{2x+1}\, dx$ with 10 subdivisions.

33. Find the area bounded between the graphs of $y = 4 - x^2$ and $y = 2x + 3$.

34. Find the area bounded by the graphs of $y = x^3$ and $y = 2x + 1$.

Chapter 5

For Problems 35 through 40, let R be the region bounded by $y = \sqrt{x+1}$, $x = 3$, and $x = 8$.

35. A solid is formed whose cross sections perpendicular to the x-axis are squares. Find the volume of the solid.

36. A solid is formed whose cross sections perpendicular to the x-axis are equilateral triangles. Find the volume of the solid.

37. A solid is formed whose cross sections perpendicular to the x-axis are semicircles. Find the volume of the solid.

38. A solid is formed whose cross sections perpendicular to the x-axis are isosceles right triangles with the hypotenuse in the plane of R. Find the volume of the solid.

39. R is rotated about the x-axis. Find the volume of the solid formed.

40. R is rotated about the y-axis. Find the volume of the solid formed.

For Problems 41 through 44, let Q be the region bounded by the graphs of $f(x) = x^3$ and $g(x) = 9x$.

41. Find the volume of the solid formed when Q is rotated about the x-axis.

42. Find the volume of the solid formed when Q is rotated about the y-axis.

43. Find the length of the arc along $f(x)$ from $x = -3$ to $x = 3$.

44. Find the surface area of the solid formed when Q is rotated about the x-axis.

Chapter 6

45. Evaluate $\int e^x (2x+1)^2 \, dx$.

46. The region bounded by the graph of $f(x) = 2\sqrt{x^3}$, the x-axis, $x = 1$, and $x = 9$ is rotated about the x-axis. Find the surface area of the solid formed.

47. Evaluate $\int x^2 \sin(2x) dx$.

48. Evaluate $\int e^{4x} \cos(5x) \, dx$.

49. Evaluate $\int \sec^7(x) \, dx$.

Chapter 7

For Problems 50 through 55, evaluate each of the integrals.

50. $\int \sqrt{81x^2 + 100} \, dx$

51. $\int \sqrt{81x^2 - 100} \, dx$

52. $\int \sqrt{100 - 81x^2} \, dx$

53. $\int \sin^3(2x) \cos^4(2x) \, dx$

54. $\int \tan^3(x) \sec^4(x) \, dx$

55. $\int \tan^4(x) \sec(x) \, dx$

Chapter 8

Evaluate each of the following integrals.

56. $\int \frac{4}{x^2-4} dx$

57. $\int \frac{x+2}{(x-1)(x-2)^2} dx$

58. $\int \frac{2x+1}{(x-4)(4x^2+9)} dx$

Chapter 9

59. Evaluate $\int_0^\infty xe^{-4x^2} dx$ if it exists.

60. Evaluate $\int_1^4 \frac{3}{2x^2-3x} dx$ if it exists.

61. Determine if $\int_1^\infty \frac{4}{\sqrt[3]{3x^2+5x+12}} dx$ is convergent or divergent.

62. R is the region bounded by $y = \frac{1}{x}$ and the x-axis on the interval $[1,\infty)$. What is the volume of the solid formed when rotated about the x-axis?

Chapter 10

63. Find the equation of the line tangent to the curve defined by $x = \cos(3t)$ and $y = \sin(5t)$ at $t = \frac{\pi}{4}$.

64. Determine the concavity of the curve defined by $x = \cos(3t)$ and $y = \sin(5t)$ at $t = \frac{\pi}{4}$.

65. Point P is located on the graph of $x = \frac{1}{2}t^2$ and $y = \frac{1}{12}(8t+17)^{3/2}$. Point P moves along the curve at a rate of 1 unit per second. How far does P move from $t = 0$ to $t = 4$?

Chapter 11

66. Find the equation of the line tangent to $r = 4\cos(3\theta)$ at $\theta = \frac{\pi}{3}$.

For Problems 67 through 69, use the information in the following graph of $r = \theta + \cos(2\theta)$ on the interval $0 \le \theta \le \pi$.

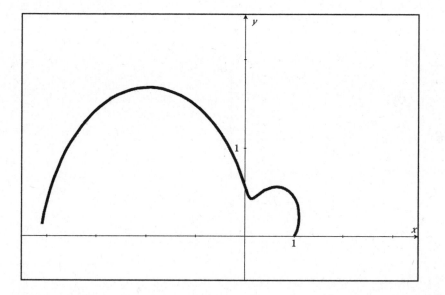

67. Write an equation for the line tangent to this curve at $\theta = \frac{\pi}{4}$.

68. Determine the length of this curve.

69. Determine the area bounded by this curve and the horizontal axis.

Chapter 12

70. Given two vectors $a = (3,6)$ and $b = (1,-5)$, find the value of $a + b$ and the measure of the angle between this sum and a.

71. An object moves in the plane with velocity vector $(4 - 3\sin(2t), 3 - \cos(t))$. Determine the velocity, speed, and acceleration of the particle at $t = \frac{5\pi}{6}$.

72. The position of a particle moving in the coordinate plane for any time on the interval $[0, 2\pi]$ is given by $x(t) = 2\cos(t) + 1$ and $y = \sin(2t)$. Determine the displacement of the particle on the interval $[0, 2\pi]$. What is the distance traveled by the particle over the same interval?

Chapter 13

73. Find the particular solution to the differential equation $\frac{dy}{dx} = \frac{\sqrt{1-y^2}}{e^x}$ if $y(0) = 0$.

74. Solve the general solution to the differential equation $\frac{dy}{dt} = 4y(80 - y)$. Consider the differential equation given by $\frac{dy}{dx} = \frac{xy}{2}$. Use this to answer questions 75 and 76.

75. On the axes provided, sketch a slope field for the given differential equation through the nine points indicated.

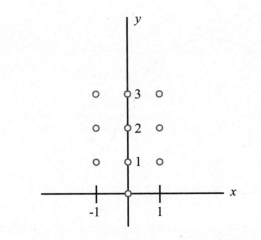

76. Let $y = f(x)$ be the particular solution to the differential equation with the initial condition $f(3) = 5$. Use Euler's Method starting at $x = 3$ with a step size of 0.1 to approximate $f(3.2)$.

77. Find the general solution $\frac{dy}{dx} + \sec(x)y = \cos(x) + 3$ for $0 < x < \frac{\pi}{2}$.

Chapter 14

78. Does the sequence $a_n = \frac{8 - 12n + 10n^3}{190 + 80n^2 - n^3}$ converge or diverge?

79. Does the sequence $a_n = \frac{\ln(n)}{n}$ converge or diverge?

80. Does the sequence $a_n = \frac{(-1)^n}{n^2 + 4}$ converge or diverge?

81. Is the sequence $\left\{ \frac{n}{10 + \sqrt{n}} \right\}$ strictly increasing, increasing, strictly decreasing, or decreasing?

Chapter 15

82. Determine whether $\sum\limits_{n=1}^{\infty} \frac{\sqrt{n}+10}{n}$ converges.

83. Determine whether $\sum\limits_{n=1}^{\infty} \frac{8-12n+10n^2}{190+80n^2-n^4}$ converges.

84. Determine whether $\sum\limits_{n=1}^{\infty} 9\sqrt[3]{n^{-2}}$ converges.

85. Determine whether $\sum\limits_{n=1}^{\infty} \frac{4^n}{n!}$ converges.

86. Determine whether $\sum\limits_{n=1}^{\infty} \frac{(-1)^n n}{e^{n^2}}$ is absolutely convergent, conditionally convergent, or divergent.

87. Determine the number of terms needed so that the maximum error for the sum

    $\sum\limits_{n=1}^{\infty} \frac{(-1)^n 4^n}{n!}$ is less than 0.001.

Chapter 16

88. Find the interval of convergence for the series $\sum\limits_{n=0}^{\infty} \frac{x^n}{n4^n}$.

89. Find the interval of convergence for the series $\sum\limits_{n=0}^{\infty} \frac{(x-5)^n}{n^2}$.

90. Express e^{-3x} as a series about $x = 0$.

91. Find the MacLaurin Series for $\sin(x^2)$.

92. Determine the first three terms for the MacLaurin Series for $\tan(x)$.

93. Determine the degree three Taylor polynomial for $\sin(x)$ about $x = \frac{\pi}{4}$.

94. What is the maximum error for computing the value of e using the first four terms of the MacLaurin Series for e^x?

Solutions

Chapter 1

1. $\frac{2\pi}{3}$

2. $-\frac{23}{32}$

3. $5x + 2\ln(\sin(x)) - 2\ln(x + 3) - \frac{1}{2}\ln(x - 1)$

4. $\frac{y^2}{9} - \frac{x^2}{25} = 1$

5. $\left(-6\sqrt{2}, -6\sqrt{2}\right)$

6. $\frac{\pi}{3}, \frac{5\pi}{3}$

7. 16

8. $\frac{6}{x+4} + \frac{5}{x+5}$

Chapter 2

9. $\frac{5}{3}$

10. 3

11. $-\sin(x)$

12. $\frac{-1}{3}$

13. $2x\ln(\cos(x)) - x^2\tan(x)$

14. $\dfrac{2x\sec^2(x^2)\sin(2x) - 2\tan(x^2)\cos(2x)}{\sin^2(2x)}$

15. $\dfrac{-\sin(x)e^{\cos(x)}}{\sqrt{1 - e^{2\cos(x)}}}$

16. $x^2\left(x^3 + 5\right)^{-\frac{2}{3}} - 3\left(6x - 2\right)^{-\frac{3}{2}}$

17. $\dfrac{dy}{dx} = -\dfrac{2xy\sec^2\left(x^2y\right) + y^2}{x^2\sec^2\left(x^2y\right) + 2xy}$

18. $\sqrt{2}e^{-1/2}$

19. (a) -5

    (b) $\frac{-9}{\sqrt{41}}$

Chapter 3

20. $f(x) = \frac{9}{5}x^5 + \frac{1}{3}(2x-3)^{3/2} + C$

21. $y = \frac{1}{8}(6x+5)^{4/3} + \frac{1}{4}e^{4x} - 3x + C$

22. $5\sin(x) + \tan^{-1}(x) - x + C$

23. $\sin^{-1}(x) - 2\sqrt{1-x} + C$

24. $g(x) = 4\ln|x+9| - \frac{1}{2}x^2 + 3x + 60$

25. $3\tan^{-1}(x) + \frac{1}{2}\ln(1+x^2) + C$

26. $\frac{\pi^2}{2}$

27. $\sin^3(x)\cos^2(x)$

28. 76.14

29. $1 - \cos(\ln(3))$

Chapter 4

30. 0.257

31. 8.671

32. 8.666

33. 3.771

34. 2.818

Chapter 5

35. $\frac{65}{2}$

36. $\frac{65\sqrt{3}}{8}$

37. $\frac{65\pi}{16}$

38. $\frac{65}{8}$

39. $\frac{65\pi}{2}$

40. 450.714 (or $\frac{2152\pi}{15}$)

41. $\frac{5832\pi}{7}$

42. $\frac{648\pi}{5}$

43. 55.316

44. 4589.637

Chapter 6

45. $4x^2 e^x - 4xe^x + 5e^x + C$

46. 9231.635

47. $\frac{-1}{2}x^2 \cos(2x) + \frac{1}{2}x\sin(2x) + \frac{1}{4}\cos(2x) + C$

48. $\frac{4}{41}e^{4x}\cos(5x) + \frac{5}{41}e^{4x}\sin(5x) + C$

49. $\frac{1}{6}\sec^5(x)\tan(x) + \frac{5}{24}\sec^3(x)\tan(x) + \frac{5}{16}\sec(x)\tan(x) + \frac{5}{16}\ln|\sec(x) + \tan(x)| + C$

Chapter 7

50. $\frac{1}{2}x\sqrt{81x^2 + 100} + \frac{50}{9}\ln\left(\sqrt{81x^2 + 100} + 9x\right) + C$

51. $\frac{1}{2}x\sqrt{81x^2 - 100} - \frac{50}{9}\ln\left(\sqrt{81x^2 - 100} + 9x\right) + C$

52. $\frac{1}{2}x\sqrt{100 - 81x^2} + \frac{50}{9}\sin^{-1}\left(\frac{9x}{10}\right) + C$

53. $\frac{1}{14}\cos^7(2x) - \frac{1}{10}\cos^5(2x) + C$

54. $\frac{1}{6}\tan^6(x) + \frac{1}{4}\tan^4(x) + C$

55. $\frac{1}{4}\sec^3(x)\tan(x) - \frac{5}{8}\sec(x)\tan(x) + \frac{3}{8}\ln|\sec(x) + \tan(x)| + C$

Chapter 8

56. $\ln\left|\frac{x-2}{x+2}\right| + C$

57. $3\ln\left|\frac{x-1}{x-2}\right| - \frac{4}{x-2} + C$

58. $\frac{9}{146}\ln\left|\frac{(x-4)^2}{4x^2+9}\right| + \frac{1}{219}\tan^{-1}\left(\frac{2x}{3}\right) + C$

Chapter 9

59. $\frac{1}{8}$

60. Limit does not exist, integral is divergent.

61. Divergent (p-test with $p = \frac{2}{3}$)

62. π

Chapter 10

63. $y + \frac{\sqrt{2}}{2} = \frac{5}{3}\left(x + \frac{\sqrt{2}}{2}\right)$

64. $\frac{20}{9\sqrt{2}}$

65. $\frac{\ln\left(\left(\sqrt{65}+8\right)\left(\sqrt{17}-4\right)\right)}{2} + 4\sqrt{65} - 2\sqrt{17}$ (or 24.3437 for those of you who use your calculator)

Chapter 11

66. $y = \frac{-\sqrt{3}}{3}(x - 4)$

67. $y - \frac{\pi\sqrt{2}}{8} = \frac{4-\pi}{\pi+4}\left(x - \frac{\pi\sqrt{2}}{8}\right)$

68. 6.954

69. $\frac{2\pi^3 + 3\pi}{12}$ (or 5.953)

Chapter 12

70. $\boldsymbol{a} + \boldsymbol{b} = (4,-1)$; 49.4°

71. Velocity $= \left(4 + \frac{3\sqrt{3}}{2}, 3 + \frac{\sqrt{3}}{2}\right)$; speed = 7.647; acceleration $= \left(-3, \frac{1}{2}\right)$

72. Displacement = 0; distance = 12.194

Chapter 13

73. $y = \sin\left(-e^{-x} + 1\right)$

74. $y = \frac{80 Ae^{320t}}{1 + Ae^{320t}}$

75.

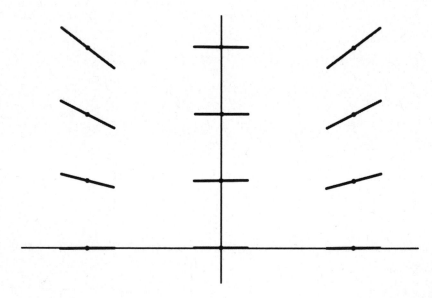

76. 6.64125

77. $y = \frac{x - \cos(x) + 3\ln\left|\sec^2(x) + \sec(x)\tan(x)\right| + C}{\sec(x) + \tan(x)}$

Chapter 14

78. Converges

79. Converges

80. Converges

81. The sequence is strictly increasing because the corresponding derivative is always positive.

Chapter 15

82. Compare with $\sum_{n=1}^{\infty} \frac{1}{\sqrt{n}}$, divergent

83. Compare with $\sum_{n=1}^{\infty} \frac{1}{n^2}$, convergent

84. p-series with $p < 1$, diverges

85. Use Ratio Test, converges

86. Absolutely convergent, use Integral Test

87. 14 terms

Chapter 16

88. $-4 \leq x < 4$

89. $4 \leq x \leq 6$

90. $\sum_{n=0}^{\infty} \frac{(-1)^n 3^n}{n!} x^n$

91. $\sum_{n=0}^{\infty} \frac{(-1)^n x^{4n+2}}{(2n+1)!}$

92. $x + \frac{x^3}{3} + \frac{2x^5}{15}$

93. $\frac{\sqrt{2}}{2} + \frac{\sqrt{2}}{2}\left(x - \frac{\pi}{4}\right) - \frac{\sqrt{2}}{4}\left(x - \frac{\pi}{4}\right)^2 - \frac{\sqrt{2}}{12}\left(x - \frac{\pi}{4}\right)^3$

94. 0.0083 (Use $e < 3$.)

Solutions to "You've Got Problems"

All the answers to the problems that you found in the "You've Got Problems" sidebars throughout the book are listed here, organized by chapter.

Chapter 1

1. Using the relationship that $180°$ corresponds to π radians, set up the proportion $\frac{30}{180} = \frac{r}{\pi}$ to arrive at $r = \frac{\pi}{6}$.

2. You know that $\sin(2A) = 2\sin(A)\cos(A)$. Use the Pythagorean identity $\sin^2(A) + \cos^2(A) = 1$ to solve for the value of $\cos(A)$. $\left(\frac{1}{\sqrt{10}}\right)^2 + \cos^2(A) = 1$ becomes $\frac{1}{10} + \cos^2(A) = 1$ so $\cos^2(A) = \frac{9}{10}$ and $\cos(A) = \frac{-3}{\sqrt{10}}$. Therefore, $\sin(2A) = 2\left(\frac{1}{\sqrt{10}}\right)\left(\frac{-3}{\sqrt{10}}\right) = \frac{-3}{5}$.

3. Separate the logarithm of a quotient into the difference of logarithms (change the square root to exponential form, too) $\ln\left(\frac{\sqrt{x+1}}{(x-2)^3}\right) = \ln\left((x+1)^{\frac{1}{2}}\right) - \ln\left((x-2)^3\right)$. Use the rule for logarithms of powers to get $\ln\left(\frac{\sqrt{x+1}}{(x-2)^3}\right) = \frac{1}{2}\ln(x+1) - 3\ln(x-2)$.

4. $\sqrt{5^2 + \left(5\sqrt{3}\right)^2} = 10$ and $\theta = \tan^{-1}\left(\frac{5\sqrt{3}}{5}\right) = \frac{\pi}{3}$ (or 1.05), so the coordinates are $\left(10, \frac{\pi}{3}\right)$.

5. The common ratio for the series is $\frac{8}{12} = \frac{\frac{16}{3}}{6} = \frac{\frac{32}{9}}{\frac{16}{3}} = \frac{2}{3}$. The first term of the series is 12, so the sum of the infinite geometric series is $S = \frac{12}{1-\frac{2}{3}} = \frac{12}{\frac{1}{3}} = 36$.

6. Factor the denominator to $(x + 5)(x - 1)$, and rewrite the fraction as $\frac{x-19}{x^2+4x-5} = \frac{A}{x+5} + \frac{B}{x-1}$. Multiply by the common denominator: $x - 19 = A(x - 1) + B(x + 5)$.

Set $x = 1$: $-18 = 6B$ so $B = -3$.

Set $x = -5$: $-24 = -6A$ so $A = 4$.

Therefore, $\frac{x-19}{x^2+4x-5} = \frac{4}{x+5} - \frac{3}{x-1}$.

Chapter 2

1. You'll get the indeterminate form $\frac{0}{0}$ when you substitute $x = -2$. Factor and reduce the fractional expression and evaluate the limit $\lim\limits_{x \to -2} \frac{-(2x-5)}{x-3}$ to get $\frac{-9}{5}$.

2. Use the product rule for the first term in the function, $p'(x) = (1)(\ln(x)) + (x)\left(\frac{1}{x}\right) - 1 = \ln(x) + 1 - 1 = \ln(x)$.

3. Find the first derivative by using the chain rule and the trigonometric identity for the sine of the double angle, $k'(x) = 2\sin(3x)\cos(3x)(3) = 3\sin(6x)$. The second derivative is also found using the chain rule, $k''(x) = 3\cos(6x)(6) = 18\cos(6x)$.

4. Evaluate the function to get the point through which the line passes, $w(2) = 5$. Find the derivative of $w(z)$ using the quotient rule, $w'(z) = \frac{(2z)(3z-5)-(z^2+1)(3)}{(3z-5)^2}$, and evaluate it at $z = 2$, $w'(2) = -11$. The equation of the line is $w - 5 = -11(z - 2)$.

5. Find the value of the first derivative. $6x - \left(x\frac{dy}{dx} + y\right) - 8y\frac{dy}{dx} = 0$ becomes

$6x - y - (x + 8y)\dfrac{dy}{dx} = 0$, so $\frac{dy}{dx} = \frac{6x-y}{x+8y}$. The derivative of this statement

is $\dfrac{d^2y}{dx^2} = \dfrac{\left[6 - \frac{dy}{dx}\right](x+8y) - (6x-y)\left[1 + 8\frac{dy}{dx}\right]}{(x+8y)^2}$. Substitute what was found for $\frac{dy}{dx}$ to get

$\dfrac{d^2y}{dx^2} = \dfrac{\left[6 - \frac{6x-y}{x+8y}\right](x+8y) - (6x-y)\left[1 + 8\left(\frac{6x-y}{x+8y}\right)\right]}{(x+8y)^2}$. Simplification beyond this is not necessary (and just plain tedious).

6. $f'(c) = 3c^2$ and $\frac{f(5)-f(1)}{5-1} = \frac{128-4}{4} = 31$, so $3c^2 = 31$ implies that $c = \sqrt{\frac{31}{3}}w$ (but not $-\sqrt{\frac{31}{3}}w$ because that is not in the interval $[1,5]$).

7. f'(x) is equal to 0 when x is approximately −6.7, 0, and approximately 3.7. f'(x) < 0 when x < −6.7 and when 0 < x < 3.7. Also, f'(x) > 0 when −6.7 < x < 0 and x > 3.7. Therefore, the graph of f(x) has relative minima at x = −6.7 and x = 3.7 and a relative maximum at x = 0.

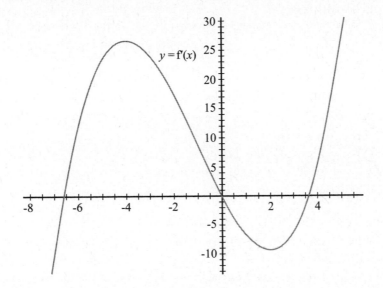

8. Plug in 0 to both the numerator and denominator to get the indeterminate form $\frac{0}{0}$. Differentiate both the numerator and denominator to get the new limit problem $\lim_{x \to 0} \frac{\sin(x)}{\cos(x)}$. Substitute 0 for x to get the answer 0.

9. The width of the rectangle is 2x, and the height of the rectangle is $4 - x^2$, so the area function is $A(x) = 8x - 2x^3$. Differentiate this function to get $A'(x) = 8 - 6x^2$. Set $A'(x) = 0$ to get $x = \frac{2}{\sqrt{3}}$. Therefore, point A has coordinates $\left(\frac{2}{\sqrt{3}}, \frac{8}{3}\right)$.

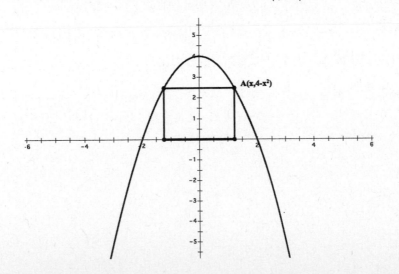

10. This is a two-part answer:

(a) When X is 9 feet from the building, Y is 12 feet above 0. $x^2 + y^2 = 225$ yields

$2x\frac{dx}{dt} + 2y\frac{dy}{dt} = 0$ or $\frac{dy}{dt} = \frac{-x}{y}\frac{dx}{dt}$. Also, $\frac{dx}{dt} = \frac{1}{2}$. $\frac{dy}{dt} = \left(\frac{-9}{12}\right)\left(\frac{1}{2}\right) = \frac{-3}{8}$ feet per second.

(b) The area of the triangle is $A = \frac{1}{2}xy$, so use the product rule to get $\frac{dA}{dt} = \frac{1}{2}x\frac{dy}{dt} + \frac{1}{2}y\frac{dx}{dt}$.

Substitute values to find that $\frac{dA}{dt} = \frac{1}{2}(9)\left(\frac{-3}{8}\right) + \frac{1}{2}(12)\left(\frac{1}{2}\right) = \frac{21}{16}$ feet squared per second.

Chapter 3

1. The antiderivative for $8x^3$ is $8\left(\frac{1}{4}x^4\right) = 2x^4$. The antiderivative for $5x$ is $5\left(\frac{1}{2}x^2\right) = \frac{5}{2}x^2$.
Rewrite $\frac{1}{\sqrt{x+2}}$ as $(x+2)^{\frac{-1}{2}}$. The antiderivative is $\frac{1}{\frac{1}{2}}(x+2)^{\frac{1}{2}} = 2\sqrt{x+2}$. Adding the
constant of integration gives $k(x) = 2x^4 + \frac{5}{2}x^2 - 2\sqrt{x+2} + C$.

2. Let $u = \sin(x^3)$ because it represents the innermost of the composed functions. $du = 3x^2\cos(x^3)dx$, so $\frac{1}{3}du = x^2\cos\left(x^3\right)dx$. The original problem $\int x^2\cos\left(x^3\right)e^{\sin\left(x^3\right)}dx$
becomes $\frac{1}{3}\int e^u\,du$ and this equals $\frac{1}{3}e^u$. Therefore, $\int x^2\cos\left(x^3\right)e^{\sin\left(x^3\right)}dx = \frac{1}{3}e^{\sin\left(x^3\right)} + C$.

3. Let $u = e^{\sin(2x)} + 1$ and $du = 2\cos(2x)e^{\sin(2x)}$, so $\frac{1}{2}du = \cos(2x)e^{\sin(2x)}$. The bounds of
integration become 2 and $e + 1$. The original problem $\int_0^{\pi/4}\frac{\cos(2x)e^{\sin(2x)}}{e^{\sin(2x)}+1}dx$ becomes
$\frac{1}{2}\int_2^{e+1}\frac{1}{u}du = \frac{1}{2}\ln|u|\Big|_2^{e+1} = \frac{1}{2}\ln(e+1) - \frac{1}{2}\ln(2)$ or $\ln\sqrt{\frac{e+1}{2}}$.

4. If $G(x)$ is the antiderivative for $\sin^3(t)$ then $f(x) = G\left(e^{x^2}\right) - G\left(\sin^{-1}(x)\right)$. This makes
$f'(x) = G'\left(e^{x^2}\right)\left(e^{x^2}\right)(2x) - G'\left(\sin^{-1}(x)\right)\left(\frac{1}{\sqrt{1-x^2}}\right) =$
$2xe^{x^2}\sin^3\left(e^{x^2}\right) - \sin^3\left(\sin^{-1}(x)\right)\left(\frac{1}{\sqrt{1-x^2}}\right) = 2xe^{x^2}\sin^3\left(e^{x^2}\right) - \frac{x^3}{\sqrt{1-x^2}}$.

Chapter 4

1. The width of each interval is $\frac{5-1}{20} = 0.2$. Use the sequence command seq($x,x,1.1,4.9,0.2$) to fill the input list.

	A in	B out
=	=seq(x,x,1.1,4.9,0.2)	=f(in)
1	1.1	4.4141
2	1.3	3.4061
3	1.5	2.8125
4	1.7	2.9021
5	1.9	3.9821
6	2.1	6.3981
7	2.3	10.5341
8	2.5	16.8125
9	2.7	25.6941
10	2.9	37.6781
11	3.1	53.3021
12	3.3	73.1421
13	3.5	97.8125
14	3.7	127.966
15	3.9	164.294
16	4.1	207.526
17	4.3	258.43
18	4.5	317.813
19	4.7	386.518
20	4.9	465.43

The estimate for the area under the curve is 453.374. (The true area is $\int_{1}^{5} x^4 - 5x^2 + 9\,dx = 454.133$.)

2. A sketch of the velocity functions shows that the object changes directions four times during this interval. It moves forward on [0,2.387] and [6,9.613], and it moves in the reverse direction on [2.387,6] and [9.613,12].

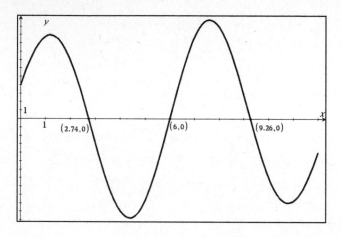

The total displacement is $\int_0^{12} 4\cos\left(\frac{\pi t}{4}\right) + 2\sin\left(\frac{\pi t}{3}\right) dt = 0$. The object returns to its starting point. The total distance traveled is $\int_0^{2.387} 4\cos\left(\frac{\pi t}{4}\right) + 2\sin\left(\frac{\pi t}{3}\right) dt +$

$\int_6^{9.613} 4\cos\left(\frac{\pi t}{4}\right) + 2\sin\left(\frac{\pi t}{3}\right) dt - \int_{2.387}^{6} 4\cos\left(\frac{\pi t}{4}\right) + 2\sin\left(\frac{\pi t}{3}\right) dt - \int_{9.613}^{12} 4\cos\left(\frac{\pi t}{4}\right) + 2\sin\left(\frac{\pi t}{3}\right) dt =$

43.382 units.

3. The graphs of the functions show that they intersect at three points, −2.21, 1.1, and 4.11. The area bounded by these two functions is $\int_{-2.21}^{1.1} f(x) - g(x)\,dx + \int_{1.1}^{4.11} g(x) - f(x)\,dx =$ 50.1501.

4. The graph of the two functions show that they intersect at +1.18. The area bounded by these two graphs is $\int_{-1.18}^{1.18} 4 - x^2 - \sec(x)\,dx = 5.105$.

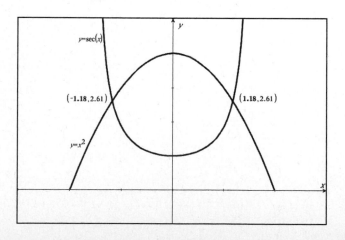

5. The width of each partition will be 0.5, and the partitions will be $[0, 0.5]$, $[0.5, 1]$, $[1, 1.5]$, $[1.5, 2]$, $[2, 2.5]$, and $[2.5, 3]$. The approximation for $\int_0^3 \sqrt{x^3 + 4}\, dx$ using Simpson's Rule is $\frac{0.5}{3}\left[\left(f(0) + 4f(0.5) + f(1)\right) + \left(f(1) + 4f(1.5) + f(2)\right) + \left(f(2) + 4f(2.5) + f(3)\right)\right] = 9.279$.

Chapter 5

1. This is a four-part answer:

   (a) Each side of the square has length $\sqrt{x + 1}$, so the volume of the solid is $\int_0^4 x + 1\, dx = 12$.

   (b) The diameter of each semicircle is $\sqrt{x + 1}$, so the volume of the solid is $\frac{\pi}{8} \int_0^4 x + 1\, dx = \frac{3\pi}{2}$.

   (c) Each side of the triangle has length $\sqrt{x + 1}$, so the volume of the solid is $\frac{\sqrt{3}}{4} \int_0^4 x + 1\, dx = 3\sqrt{3}$.

   (d) Each leg of the right triangle has length $\sqrt{\dfrac{x + 1}{2}}$, so the volume of the solid is $\frac{1}{4} \int_0^4 x + 1\, dx = 3$.

2. The cross section of this solid of revolution will be a washer. The radius of the larger circle is $\left(4 - \frac{4}{3}x\right) + 2$ or $6 - \frac{4}{3}x$, and the radius of the smaller circle is 2. The volume of the solid formed is $\pi \int_0^3 \left(6 - \frac{4}{3}x\right)^2 - 4\, dx = 40\pi$.

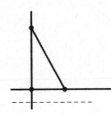

3. When the region R is rotated around the x-axis, any cross section of the solid formed will be a disk with radius $\frac{1}{\sqrt{1+x^2}}$. Therefore, the volume of the solid formed will be

$$\pi \int_0^1 \frac{1}{1+x^2}\,dx = \pi \tan^{-1}(x)\Big|_0^1 = \frac{\pi^2}{4}.$$

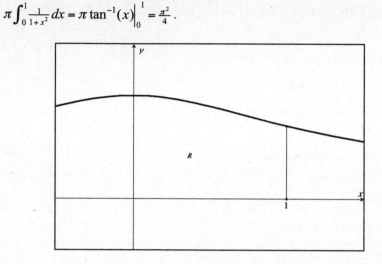

4. Use the cylindrical shell method with radius $= x$ and height $= (4x - x^2) - 2x = 2x - x^2$. The volume of the solid formed is $2\pi \int_0^2 x(2x - x^2)\,dx = y^2 = 16 - \frac{16x^2}{25}.$

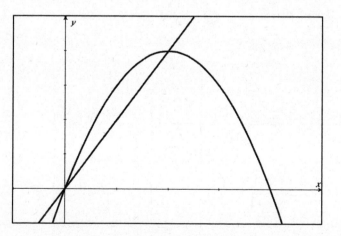

5. Rewrite the equation as $y^2 = 16 - \frac{16x^2}{25}$ so that $2y\frac{dy}{dx} = \frac{-32x}{25}$ making $\frac{dy}{dx} = \frac{-16x}{25\sqrt{16 - \frac{16x^2}{25}}}$.

Simplify this fraction to be $\frac{dy}{dx} = \frac{-16x}{5\sqrt{400 - 16x^2}}$. This makes $\left(\frac{dy}{dx}\right)^2 = \frac{256x^2}{25(400 - 16x^2)}$. The length of the arc in the first quadrant from $x = 0$ to $x = 5$ is $\int_0^5 \sqrt{1 + \frac{256x^2}{25(400 - 16x^2)}}\,dx = 7.0904.$

6. $\frac{dy}{dx} = 4 - 2x$ giving the surface area $2\pi \int_0^4 \left(4x - x^2\right)\sqrt{1 + (4 - 2x)^2}\, dx = 127.121$ (which is

an approximation for the exact answer $\dfrac{\left(\left(65\ln\left(\sqrt{17} + 4\right)\right) - 65\left(\ln\left(\sqrt{17} - 4\right)\right) + 248\sqrt{17}\right)\pi}{32}$).

Chapter 6

1. Let $u = x$ and $dv = 36(9x + 1)^{\frac{1}{3}}\, dx$. Then $du = dx$ and $v = 3(9x + 1)^{\frac{4}{3}}$. $\int 36x\sqrt[3]{9x + 1}\, dx$

$= 3x(9x + 1)^{\frac{4}{3}} - 3\int (9x + 1)^{\frac{4}{3}}\, dx = 3x(9x + 1)^{\frac{4}{3}} - \frac{1}{7}\left(9x + 1\right)^{\frac{7}{3}} + C$.

2. Solution 1 (by formula):

Let $u = x^2$ and $dv = \sqrt{8x + 3}\,dx$ This makes $du = 2x\,dx$ and $v = \frac{1}{12}\left(8x + 3\right)^{\frac{3}{2}}$.

$\int x^2\sqrt{8x + 3}\,dx = \frac{1}{12}x^2\left(8x + 3\right)^{\frac{3}{2}} - \frac{1}{6}\int x\left(8x + 3\right)^{\frac{3}{2}}\, dx$

Let $u = x$ and $dv = \left(8x + 3\right)^{\frac{3}{2}}\, dx$. This makes $du = dx$ and $v = \frac{1}{20}\left(8x + 3\right)^{\frac{5}{2}}$.

$\int x^2\sqrt{8x + 3}\,dx = \frac{1}{12}x^2\left(8x + 3\right)^{\frac{3}{2}} - \frac{1}{6}\left(\frac{1}{20}x(8x + 3)^{\frac{5}{2}} - \frac{1}{20}\int (8x + 3)^{\frac{5}{2}}\, dx\right)$

$\int x^2\sqrt{8x + 3}\,dx = \frac{1}{12}x^2\left(8x + 3\right)^{\frac{3}{2}} - \frac{1}{120}x(8x + 3)^{\frac{5}{2}} + \frac{1}{120}\int (8x + 3)^{\frac{5}{2}}\, dx$

$\int x^2\sqrt{8x + 3}\,dx = \frac{1}{12}x^2\left(8x + 3\right)^{\frac{3}{2}} - \frac{1}{120}x(8x + 3)^{\frac{5}{2}} + \frac{1}{3360}(8x + 3)^{\frac{7}{2}} + C$

Solution 2 (tabular):

Let $u = x^2$ and $dv = \sqrt{8x + 3}\,dx$.

u	dv	$+1$
x^2	$\sqrt{8x + 3}\,dx$	1
$2x$	$\frac{1}{12}\left(8x + 3\right)^{\frac{3}{2}}$	-1
2	$\frac{1}{240}\left(8x + 3\right)^{\frac{5}{2}}$	1
0	$\frac{1}{6720}\left(8x + 3\right)^{\frac{7}{2}}$	-1
		1

$\int x^2\sqrt{8x + 3}\,dx = \frac{1}{12}x^2\left(8x + 3\right)^{\frac{3}{2}}(1) + \frac{1}{240}(2x)\left(8x + 3\right)^{\frac{5}{2}}(-1) + \frac{2}{6720}\left(8x + 3\right)^{\frac{7}{2}}(1) + C$

$\int x^2\sqrt{8x + 3}\,dx = \frac{1}{12}x^2\left(8x + 3\right)^{\frac{3}{2}} - \frac{1}{120}x(8x + 3)^{\frac{5}{2}} + \frac{1}{3360}(8x + 3)^{\frac{7}{2}} + C$

3. Let $u = \ln(x)$ and $dv = \sqrt{x}dx$. This gives $du = \frac{1}{x}dx$ and $v = \frac{2}{3}x^{3/2}$.

$$\int \sqrt{x}\ln(x)\,dx = \left(\frac{2}{3}x^{3/2}\right)\ln(x) - \frac{2}{3}\int\left(\frac{1}{x}\right)\left(x^{3/2}\right)dx = \frac{2}{3}x^{3/2}\ln(x) - \frac{2}{3}\int x^{1/2}\,dx =$$

$$\frac{2}{3}x^{3/2}\ln(x) - \frac{2}{3}\left(\frac{2}{3}x^{3/2}\right) + C$$

$$\int \sqrt{x}\ln(x)\,dx = \frac{2}{3}x^{3/2}\ln(x) - \frac{4}{9}x^{3/2} + C$$

4. If $u = e^{2x}$ and $dv = \cos(3x)dx$, then $du = 2e^{2x}dx$ and $v = \frac{1}{3}\sin(3x)$.

$$\int e^{2x}\cos(3x)\,dx = \frac{1}{3}e^{2x}\sin(3x) - \frac{2}{3}\int e^{2x}\sin(3x)\,dx$$

A second application of integration by parts with $u = e^{2x}$ and $dv = \sin(3x)dx$ gives $du = 2e^{2x}dx$ and $v = \frac{-1}{3}\cos(x)$.

$$\int e^{2x}\cos(3x)\,dx = \frac{1}{3}e^{2x}\sin(3x) - \frac{2}{3}\left(\frac{-1}{3}e^{2x}\cos(3x) - \frac{-2}{3}\int e^{2x}\cos(3x)\,dx\right) =$$

$$\frac{1}{3}e^{2x}\sin(3x) + \frac{2}{9}e^{2x}\cos(3x) - \frac{4}{9}\int e^{2x}\cos(3x)\,dx$$

Add $\frac{4}{9}\int e^{2x}\cos(3x)\,dx$ to both sides of the equation to get $\frac{13}{9}\int e^{2x}\cos(3x)\,dx =$

$\frac{1}{3}e^{2x}\sin(3x) + \frac{2}{9}e^{2x}\cos(3x)$. Multiply both sides of the equation by $\frac{9}{13}$.

$$\int e^{2x}\cos(3x)\,dx = \frac{9}{13}\left(\frac{1}{3}e^{2x}\sin(3x) + \frac{2}{9}e^{2x}\cos(3x)\right) = \frac{3}{13}e^{2x}\sin(3x) + \frac{2}{13}e^{2x}\cos(3x) + C$$

Chapter 7

1. Let $5x = 7\tan(\theta)$ so that $dx = \frac{7}{5}\sec^2(\theta)d\theta$ and $\sqrt{49 + 25x^2} = 7\sec(\theta)$.

$\int \sqrt{49 + 25x^2}\,dx$ becomes $\int\left(7\sec(\theta)\right)\left(\frac{7}{5}\sec^2(\theta)d\theta\right) = \frac{49}{5}\int \sec^3(\theta)\,d\theta = \frac{49}{5}\left(\frac{1}{2}\sec(\theta)\tan(\theta) + \frac{1}{2}\ln|\sec(\theta) + \tan(\theta)|\right)$.

Return to the original variable and include the bounds of integration.

$$\int \sqrt{49 + 25x^2}\,dx = \frac{49}{10}\left(\left(\frac{\sqrt{49+25x^2}}{7}\right)\left(\frac{5x}{7}\right) + \ln\left|\left(\frac{\sqrt{49+25x^2}}{7}\right) + \left(\frac{5x}{7}\right)\right|\right)\Bigg|_0^{10} =$$

$$\frac{49}{10}\left[\left(\left(\frac{\sqrt{2549}}{7}\right)\left(\frac{50}{7}\right) + \ln\left|\frac{\sqrt{2549}}{7} + \frac{50}{7}\right|\right) - \left((1)(0) + \ln|1 + 0|\right)\right]$$

$$= 5\sqrt{2549} + \frac{49}{10}\ln\left|\frac{50 + \sqrt{2549}}{7}\right|$$

2. Let $3x = 5\sec(\theta)$ so that $dx = \frac{5}{3}\sec(\theta)\tan(\theta)d\theta$ and $\sqrt{9x^2 - 25} = 5\tan(\theta)$.

$\int \sqrt{9x^2 - 25}\, dx$ becomes $\frac{25}{3}\int \sec(\theta)\tan^2(\theta)d\theta = \frac{25}{3}\int \sec^3(\theta) - \sec(\theta)d\theta =$
$\frac{25}{3}\left(\frac{1}{2}\sec(\theta)\tan(\theta) - \frac{1}{2}\ln|\sec(\theta) + \tan(\theta)|\right) + C$

Return to the original variable.

$\int \sqrt{9x^2 - 25}\, dx = \frac{1}{2}x\sqrt{9x^2 - 25} - \frac{25}{6}\ln|3x + \sqrt{9x^2 - 25}| + C$

3. Let $\sin(\theta) = \frac{10x}{7}$ so that $dx = \frac{7}{10}\cos(\theta)d\theta$ and $\sqrt{49 - 100x^2} = 7\cos(\theta)$.

$\int \sqrt{49 - 100x^2}\, dx$ becomes $\frac{49}{10}\int \cos^2(\theta)d\theta = \frac{49}{10}\int \frac{\cos(2\theta) + 1}{2}d\theta = \frac{49}{20}\int \cos(2\theta) + 1 d\theta =$
$\frac{49}{20}\left(\frac{1}{2}\sin(2\theta) + \theta\right) = \frac{49}{20}\left(\left(\frac{1}{2}\right)2\sin(\theta)\cos(\theta) + \theta\right)$.

Transform this result back to the original variable to get $\int \sqrt{49 - 100x^2}\, dx =$
$\frac{49}{20}\left(\left(\frac{10x}{7}\right)\left(\frac{\sqrt{49 - 100x^2}}{7}\right) + \sin^{-1}\left(\frac{10x}{7}\right)\right) + C = \frac{1}{2}x\sqrt{49 - 100x^2} + \frac{49}{20}\sin^{-1}\left(\frac{10x}{7}\right) + C$.

4. Rewrite $\cos^9(x)$ as $\cos^8(x)\cos(x)$ and then as $(1 - \sin^2(x))^4\cos(x)$. The integral now becomes
$\int \sin^9(x)\left(1 - \sin^2(x)\right)^4 \cos(x)dx$. Expand the binomial and then distribute the factor
$\sin^9(x) = \cos(x)$ through the result.

$\int \sin^9(x)\left(1 - 4\sin^2(x) + 6\sin^4(x) - 4\sin^6(x) + \sin^8(x)\right)\cos(x)dx$

$\int \sin^9(x)\cos(x) - 4\sin^{11}(x)\cos(x) + 6\sin^{13}(x)\cos(x) - 4\sin^{15}(x)\cos(x) + \sin^{17}(x)\cos(x)dx$

$\int \sin^9(x)\cos^9(x)dx = \frac{1}{10}\sin^{10}(x) - \frac{1}{3}\sin^{12}(x) + \frac{3}{7}\sin^{14}(x) - \frac{1}{4}\sin^{16}(x) + \frac{1}{18}\sin^{18}(x) + C$

5. Rewrite $\sec^{10}(5x)$ as $\sec^8(5x)\sec^2(5x)$ and then $(\sec^2(5x))^4$. Use the Pythagorean identity on this 4th powered term to get $(1 + \tan^2(5x))^4$.

$\int \tan^2(5x)\sec^{10}(5x)dx = \int \tan^2(5x)\sec^2(5x)\left(1 + \tan^2(5x)\right)^4 dx$

$= \int \tan^2(5x)\sec^2(5x)\left(1 + 4\tan^2(5x) + 6\tan^4(5x) + 4\tan^6(5x) + \tan^8(5x)\right)dx$

$= \int \tan^2(5x)\sec^2(5x) + 4\tan^4(5x)\sec^2(5x) + 6\tan^6(5x)\sec^2(5x) + 4\tan^8(5x)\sec^2(5x) +$
$\tan^{10}(5x)\sec^2(5x)dx$

$= \frac{1}{15}\tan^3(5x) + \frac{4}{25}\tan^5(5x) + \frac{6}{35}\tan^7(5x) + \frac{4}{45}\tan^9(5x) + \frac{1}{55}\tan^{11}(5x) + C$

6. Rewrite $\tan^2(x)$ as $\sec^2(x) - 1$. This makes $\int \tan^2(x)\sec^3(x)dx = \int \sec^5(x) - \sec^3(x)dx$.

$= \frac{1}{4}\sec^3(x)\tan(x) + \frac{3}{8}\sec(x)\tan(x) + \frac{3}{8}\ln|\sec(x) + \tan(x)| +$
$\frac{1}{2}\sec(x)\tan(x) + \frac{1}{2}\ln|\sec(x) + \tan(x)| + C$

$= \frac{1}{4}\sec^3(x)\tan(x) + \frac{7}{8}\sec(x)\tan(x) + \frac{7}{8}\ln|\sec(x) + \tan(x)| + C$

Chapter 8

1. Complete the square: $9x^2 + 12x + 20 = 9\left(x^2 + \frac{4}{3}x + \left(\frac{2}{3}\right)^2\right) + 20 - 9\left(\frac{2}{3}\right)^2 = 9\left(x + \frac{2}{3}\right)^2 + 16$

   $= (3x + 2)^2 + 16 = 16\left(1 + \left(\frac{3x+2}{4}\right)^2\right)$. $\int \frac{1}{9x^2 + 12x + 20}dx = \frac{1}{16}\int \frac{1}{1+\left(\frac{3x+2}{4}\right)^2}dx$. Let $u = \frac{3x+2}{4}$ so

   that $du = \frac{3}{4}dx$ or $dx = \frac{4}{3}du$. $\frac{1}{12}\int \frac{1}{1+u^2}du = \frac{1}{12}\tan^{-1}(u)$. Return to the original variable,

   $\int \frac{1}{9x^2 + 12x + 20}dx = \frac{1}{12}\tan^{-1}\left(\frac{3x+2}{4}\right) + C$

2. Complete the square: $-4x^2 - 6x - 2 = -4\left(x^2 + \frac{3}{2}x + \left(\frac{3}{4}\right)^2\right) - 2 + 4\left(\frac{3}{4}\right)^2 =$

   $\frac{1}{4} - 4\left(x + \frac{3}{4}\right)^2 = \frac{1}{4}\left(1 - 16\left(x + \frac{3}{4}\right)^2\right) = \frac{1}{4}\left(1 - \left(4x + 3\right)^2\right)$. Then $\int \frac{1}{\sqrt{-4x^2 - 6x - 2}}dx =$

   $\int \frac{1}{\sqrt{\frac{1}{4}\left(1 - (4x+3)^2\right)}}dx = \int \frac{1}{\frac{1}{2}\sqrt{\left(1 - (4x+3)^2\right)}}dx = 2\int \frac{1}{\sqrt{1 - (4x+3)^2}}dx$. Let $u = 4x + 3$ so $du = 4dx$ or dx

   $= \frac{1}{4}du$. $2\int \frac{1}{\sqrt{1 - (4x+3)^2}}dx = \frac{2}{4}\int \frac{1}{\sqrt{1-u^2}}du = \frac{1}{2}\sin^{-1}(u)$. Returning to the original variable,

   $\int \frac{1}{\sqrt{-4x^2 - 6x - 2}}dx = \frac{1}{2}\sin^{-1}(4x + 3) + C$.

3. The factors of $3x^2 + 10x + 8$ are $(3x + 4)$ and $(x + 2)$. The equation for decomposing the
   fraction is $\frac{1}{3x^2 + 10x + 8} = \frac{A}{3x+4} + \frac{B}{x+2}$.

   Multiply through by the common denominator: $1 = A(x + 2) + B(3x + 4)$.

   Set $x = -2$: $1 = B(-2)$ so $B = \frac{-1}{2}$.

   Set $x = \frac{-4}{3}$: $1 = A\left(\frac{2}{3}\right)$ so $A = \frac{3}{2}$.

   $\int \frac{1}{3x^2 + 10x + 8}dx = \frac{3}{2}\int \frac{1}{3x+4}dx + \frac{1}{2}\int \frac{1}{x+2}dx = \frac{1}{2}\ln|3x + 4| - \frac{1}{2}\ln|x + 2| + C = \frac{1}{2}\ln\left|\frac{3x+4}{x+2}\right| + C$

4. The denominator is already factored, so write the equation
   $\frac{2x+3}{(x+1)(3x-2)^2} = \frac{A}{x+1} + \frac{B}{3x-2} + \frac{C}{(3x-2)^2}$.

   Multiply through by the common denominator.

   $2x + 3 = A(3x - 2)^2 + B(x + 1)(3x - 2) + C(x + 1)$

   Set $x = \frac{2}{3}$: $2\left(\frac{2}{3}\right) + 3 = C\left(\frac{5}{3}\right)$, which becomes $\frac{13}{3} = \frac{5}{3}C$ so $C = \frac{13}{5}$.

   Set $x = -1$: $2(-1) + 3 = A(3(-1) - 2)^2$, which becomes $\frac{1}{25} = A$.

   Set $x = 0$: $3 = \frac{1}{25}(-2)^2 + B(1)(-2) + \frac{13}{5}(1)$ or $3 = \frac{69}{25} - 2B$ and $B = \frac{-3}{25}$.

   $\int \frac{2x+3}{(x+1)(3x-2)^2}dx = \frac{1}{25}\int \frac{1}{x+1}dx - \frac{3}{25}\int \frac{1}{3x-2}dx + \frac{13}{5}\int \frac{1}{(3x-2)^2}dx = \frac{1}{25}\ln|x + 1| - \frac{1}{25}\ln|3x - 2| -$

   $\frac{13}{15}\left(\frac{1}{3x-2}\right) + C$

   $\int \frac{2x+3}{(x+1)(3x-2)^2}dx = \frac{1}{25}\ln|x + 1| - \frac{1}{25}\ln|3x - 2| - \frac{13}{15}\left(\frac{1}{3x-2}\right) + C$

5. The factors of $x^3 + 4x$ are x and $(x^2 + 4)$. The equation for decomposing the fraction is $\frac{5}{x^3 + 4x} = \frac{A}{x} + \frac{Bx + C}{x^2 + 4}$.

Multiply through by the common denominator:

$5 = A(x^2 + 4) + (Bx + C)(x)$

Set $x = 0$: $5 = 4A$ so $A = \frac{5}{4}$.

We cannot eliminate any other terms with a convenient selection of x, so we will have to solve a system of equations.

Set $x = 1$ (yes, an easy number again): $5 = \frac{5}{4}(5) + (B + C)$ so $B + C = \frac{-5}{4}$.

Set $x = -1$ (or whatever number you like): $5 = \frac{5}{4}(5) + (-B - C)$ so $-B - C = \frac{-5}{4}$.

Add the two equations to get $2B = \frac{-5}{2}$ so $B = \frac{-5}{4}$.

If $B + C = \frac{-5}{4}$ and $B = \frac{-5}{4}$, then $C = 0$.

$\int \frac{5}{x^3 + 4x}\,dx = \frac{5}{4}\int \frac{1}{x}\,dx - \frac{5}{4}\int \frac{x}{x^2 + 4}\,dx = \frac{5}{4}\ln|x| - \frac{5}{8}\ln\left|x^2 + 4\right| + C$

Chapter 9

1. $\int_{-\infty}^{\infty} \frac{1}{e^{-x} + e^x}\,dx = \int_{-\infty}^{\infty}\left(\frac{1}{e^{-x} + e^x}\right)\frac{e^x}{e^x}\,dx = \int_{-\infty}^{\infty}\frac{e^x}{1 + e^{2x}}\,dx$. Rewrite $\int_{-\infty}^{\infty}\frac{e^x}{1 + e^{2x}}\,dx$ as $\int_{-\infty}^{0}\frac{e^x}{1 + e^{2x}}\,dx +$

$\int_{0}^{\infty}\frac{e^x}{1 + e^{2x}}\,dx$. The antiderivative of $\frac{e^x}{1 + e^{2x}}$ is $\tan^{-1}(e^x)$. Letting $u = e^x$ and $du = e^x\,dx$ gives

$\frac{1}{1 + u^2}$.

$\int_{-\infty}^{0}\frac{e^x}{1 + e^{2x}}\,dx = \lim_{n \to -\infty}\int_{n}^{0}\frac{e^x}{1 + e^{2x}}\,dx = \lim_{n \to -\infty}\left(\tan^{-1}(e^x)\Big|_{n}^{0}\right) = \lim_{n \to -\infty}\left(\tan^{-1}(1) - \tan^{-1}(e^n)\right) =$

$\frac{\pi}{4} - 0 = \frac{\pi}{4}$

$\int_{0}^{\infty}\frac{e^x}{1 + e^{2x}}\,dx = \lim_{n \to \infty}\int_{0}^{n}\frac{e^x}{1 + e^{2x}}\,dx = \lim_{n \to \infty}\left(\tan^{-1}(e^x)\Big|_{0}^{n}\right) = \lim_{n \to \infty}\left(\tan^{-1}(e^n) - \tan^{-1}(1)\right) =$

$\frac{\pi}{2} - \frac{\pi}{4} = \frac{\pi}{4}$

Therefore, $\int_{-\infty}^{\infty}\frac{1}{e^{-x} + e^x}\,dx = \frac{\pi}{4} + \frac{\pi}{4} = \frac{\pi}{2}$.

2. $\int_{-5}^{0}\frac{1}{\sqrt{25 - x^2}}\,dx$ becomes $\lim_{n \to -5^+}\int_{n}^{0}\frac{1}{\sqrt{25 - x^2}}\,dx = \lim_{n \to -5^+}\int_{n}^{0}\frac{1}{5\sqrt{1 - \left(\frac{x}{5}\right)^2}}\,dx = \lim_{n \to -5^+}\left(\sin^{-1}\left(\frac{x}{5}\right)\Big|_{n}^{0}\right) =$

$\lim_{n \to -5^+}\left(\sin^{-1}(0) - \sin^{-1}\left(\frac{n}{5}\right)\right) = \frac{\pi}{2}$.

3. Because $\frac{2}{\sqrt[3]{x^4}} \geq \frac{2}{\sqrt[3]{x^4 + 3x^2 + 1}} \geq 0$ and $\sqrt[3]{x^4} = x^{4/3}$ making $\frac{2}{\sqrt[3]{x^4}}$ by the p-test, $\int_{2}^{\infty}\frac{2}{\sqrt[3]{x^4 + 3x^2 + 1}}\,dx$ converges.

Chapter 10

1. We'll need to find the point of tangency as well as the slope of the tangent line. At $t = 0$, $x = e^0\tan(0) = 0$ while $y = e^0\sec(0) = 1$. To find the slope, we need to find the value of $\sqrt{\frac{31}{3}}w$. We compute $\frac{dx}{dt}$ and $\frac{dy}{dt}$, $\frac{dx}{dt} = e^t\tan(2t) + 2e^t\sec^2(2t)$ and $\frac{dy}{dt} = 2e^{2t}\sec(t) + e^{2t}\sec(t)\tan(t)$. At $t = 0$, $\frac{dy}{dx} = \frac{2e^0\sec(0) + e^0\sec(0)\tan(0)}{e^0\tan(0) + 2e^0\sec^2(0)} = \frac{2}{2} = 1$. The equation of the tangent line is $y = x + 1$.

2. The first derivative is $\frac{dy}{dx} = \frac{\frac{dy}{dt}}{\frac{dx}{dt}} = \frac{12\cos(4t)}{-12\sin(3t)} = \frac{-\cos(4t)}{\sin(3t)}$. The second derivative is $\frac{d\left(\frac{\left(\frac{dy}{dx}\right)}{dt}\right)}{\frac{dx}{dt}} = \frac{\frac{\sin(3t)(4\sin(4t)) - (-\cos(4t))(3\cos(3t))}{\sin^2(3t)}}{-12\sin(3t)}$. Rather than trying to simplify the fraction, we'll evaluate it at

$$t = \frac{\pi}{2}. \frac{d^2y}{dx^2} = \frac{\frac{\sin\left(\frac{3\pi}{2}\right)(4\sin(2\pi)) - \left(-\cos(2\pi)\right)\left(3\cos\left(\frac{3\pi}{2}\right)\right)}{\sin^2\left(\frac{3\pi}{2}\right)}}{-12\sin\left(\frac{3\pi}{2}\right)} = \frac{\frac{(-1)(0) - (-1)(-3)(0)}{1}}{12} = 0$$

3. $\frac{dx}{dt} = \frac{1}{1+t^2}$ and since y can be rewritten as $y = \frac{1}{2}\ln\left(1 + t^2\right)$ $\frac{dy}{dt} = \frac{t}{1+t^2}$. The length of the arc is $\int_0^1\sqrt{\left(\frac{1}{1+t^2}\right)^2 + \left(\frac{t}{1+t^2}\right)^2}\,dt = \int_0^1\sqrt{\frac{1}{\left(1+t^2\right)^2} + \frac{t^2}{\left(1+t^2\right)^2}}\,dt = \int_0^1\sqrt{\frac{1+t^2}{\left(1+t^2\right)^2}}\,dt = \int_0^1\sqrt{\frac{1}{1+t^2}}\,dt = \int_0^1\frac{1}{\sqrt{1+t^2}}\,dt$. The sum of squares in the denominator leads us to consider trigonometric substitution.

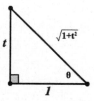

$\frac{1}{\sqrt{1+t^2}} = \cos(\theta)$, $t = \tan(\theta)$ so $dt = \sec^2(\theta)d\theta$.

$\int\frac{1}{\sqrt{1+t^2}}\,dt$ becomes $\int\cos(\theta)\sec^2(\theta)d\theta = \int\sec(\theta)d\theta = \ln|\sec(\theta) + \tan(\theta)|$. Returning to the original variable, $\int_0^1\frac{1}{\sqrt{1+t^2}}\,dt = \ln\left|\sqrt{1+t^2} + t\right|\Big|_0^1 = \ln\left|\sqrt{2} + 1\right|$.

Chapter 11

1. The point is at $x = r\cos(\theta) = \frac{-11}{4}$ and $y = r\sin(\theta) = \frac{-11\sqrt{3}}{4}$. The slope of the tangent line is $\frac{dy}{dx} = \frac{\frac{d(r\sin(\theta))}{d\theta}}{\frac{d(r\cos(\theta))}{d\theta}} = \frac{5\sin^2(\theta) + (3 - 5\cos(\theta))\cos(\theta)}{5\sin(\theta)\cos(\theta) - (3 - 5\cos(\theta))\sin(\theta)}$. At $\theta = \frac{2\pi}{3}$, $\frac{dy}{dx} = \frac{-1}{4\sqrt{3}}$, so the equation of the tangent lines is $y - \frac{11\sqrt{3}}{4} = \frac{-1}{4\sqrt{3}}\left(x + \frac{11}{4}\right)$.

2. There are three petals to the rose $r = 4\sin(3\theta)$. The first petal begins at $\theta = 0$ and reaches its tip at $\theta = \frac{\pi}{6}$. The full distance around the rose is equal to six times the distance along the petal from $\theta = 0$ to $\theta = \frac{\pi}{6}$. The total distance is

$$6\int_0^{\pi/6} \sqrt{\left(4\sin(3\theta)\right)^2 + \left(12\cos(3\theta)\right)^2}\, d\theta = 26.730.$$

3. As you did with the distance around the rose, find the area of the figure from $\theta = 0$ to $\theta = \frac{\pi}{6}$ and multiply by 6. $\frac{1}{2}\int_0^{\pi/6} 16\sin^2(3\theta)d\theta = 4\pi.$

Chapter 12

1. The sum is $a + \mathbf{b} = (2,19)$. The angle between a and $a + b$ is given by $\cos(\theta) = \frac{181}{\left(\sqrt{106}\right)\left(\sqrt{365}\right)}$ so $\theta = 23.05°$.

2. The acceleration vector is $(-\cos(t), 2\sin(t))$. Velocity $= \left(3 - \frac{\sqrt{3}}{2}, 3\right)$; speed $= 3.682$; acceleration $= \left(\frac{-1}{2}, \sqrt{3}\right)$.

3. The velocity vector is $(2\cos(t), -2\sin(2t))$.

   (a) Velocity $= \left(-1, \sqrt{3}\right)$; speed $= 2$.

   (b) The velocity is equal to 0 when $t = \frac{\pi}{2}, \frac{3\pi}{2}$.

   (c) Displacement $(0,0)$; distance $= 11.832$.

Chapter 13

1. $\int \frac{\ln(y)}{y}\, dy = \int x\, dx$ becomes $\frac{1}{2}\left(\ln(y)\right)^2 = \frac{1}{2}x^2 + C$ so that $(\ln(y))^2 = x^2 + C$. Use the given point to get $(\ln(y))^2 = 16 - x^2$.

2. (a) $\frac{dr}{dt} = kr$; $r = Ae^{kt}$.

   (b) At $t = 0$, $4 \times 10^{18} = Ae^{k(0)}$ so $A = 4 \times 10^{18}$. Solve $5 \times 10^{14} = 4 \times 10^{18}\, e^{k(90)}$ to get $r = (4 \times 10^{18})e^{-0.0998577t}$.

   (c) Solve $4 \times 10^{18}\, e^{kt} = 8 \times 10^{-4}$; $t = 500.3$ seconds.

3. $f(3.2) = 10.192$

4.

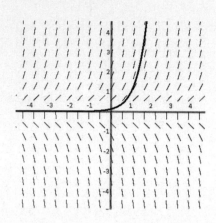

5. The integration factor is $e^{\int 2\,dx} = e^{2x}$, so the solution is $y = \frac{1}{2}x + \frac{5}{4} + Ce^{-2x}$.

Chapter 14

1. $\lim\limits_{n \to \infty} \dfrac{n^2 + n + 100000}{2n^2 - 1} = \lim\limits_{n \to \infty} \dfrac{1 + \frac{1}{n} + \frac{100000}{n^2}}{2 - \frac{1}{n}} = \frac{1}{2}$. The sequence converges.

2. We showed in Example 4 that $\left\{\dfrac{n}{n+1}\right\}$ converges, and this is $\left|\dfrac{(-1)^n n}{n+1}\right|$. Because $\left|\dfrac{(-1)^n n}{n+1}\right|$ converges, so does $\left\{\dfrac{(-1)^n n}{n+1}\right\}$.

3. Use the function $f(x) = xe^{-2x}$ and its derivative $f'(x) = e^{-2x} - 2xe^{-2x} = e^{-2x}(1 - 2x)$. Because $f'(x) < 0$ whenever $x > \frac{1}{2}$, the sequence is strictly decreasing.

Chapter 15

1. $S = \dfrac{36}{1 - \left(\frac{-2}{3}\right)} = \dfrac{108}{5}$

2. $\lim\limits_{p \to \infty} \int_1^p \dfrac{x}{e^x}\,dx = \lim\limits_{p \to \infty} \int_1^p xe^{-x}\,dx$. Use integration by parts to get $\lim\limits_{p \to \infty} -e^{-x}(x+1)\Big|_1^p = \lim\limits_{p \to \infty} -e^{-p}(p+1) + e^{-1}(2) = \dfrac{2}{e}$. Therefore, $\sum\limits_{n=1}^{\infty} \dfrac{n}{e^n}$ converges.

3. Compare $\sum \dfrac{n+3}{n(n+1)(n+2)}$ with $\sum \dfrac{1}{n^2}$. $\dfrac{1}{n^2} > \dfrac{n+3}{n^3 + 3n^2 + 2n}$, which becomes $n^3 + 3n^2 + 2n^2 > n^3 + 3n$. Solve $2n > 0$ to determine the inequality is true when $n > 0$. The series $\sum \dfrac{n+3}{n(n+1)(n+2)}$ converges.

4. Compare this series with $\sum_{k=1}^{\infty} \frac{1}{k^{1/3}}$. $\lim_{k \to \infty} \frac{\frac{8k^2}{\sqrt[3]{4k^7+9k^3}}}{\frac{1}{k^{1/3}}} = \lim_{k \to \infty} \frac{8k^2}{\sqrt[3]{4k^7+9k^3}} \times k^{1/3} = \lim_{k \to \infty} \frac{8k^{7/3}}{\sqrt[3]{4k^7+9k^3}} = 2.$

Since $\sum_{k=1}^{\infty} \frac{1}{k^{1/3}}$ diverges, so does $\sum_{k=1}^{\infty} \frac{8k^2}{\sqrt[3]{4k^7+9k^3}}$.

5. The term $\sin\left(\frac{(2n-1)\pi}{2}\right)$ is the same as $(-1)^n$, so the series $\sum \frac{\sin\left(\frac{(2n-1)\pi}{2}\right)}{\sqrt{n}}$ is the same as $\sum \frac{(-1)^n}{\sqrt{n}}$. The series meets both conditions for the Alternating Series Test, so $\sum \frac{\sin\left(\frac{(2n-1)\pi}{2}\right)}{\sqrt{n}}$ converges.

6. Use the Ratio test: $\lim_{n \to \infty} \left| \frac{(n+1)4^{n+1}}{(n+1)!} \times \frac{n!}{n4^n} \right| = \lim_{n \to \infty} \left| \frac{4}{n} \right| = 0$. Therefore, the series $\sum_{n=1}^{\infty} \frac{n4^n}{n!}$ is absolutely convergent.

7. We need $|S - s_n| < a_{n+1} < 0.00001$. This becomes $\frac{(n+1)^2}{2^{n+1}} < 0.00001$. Use the List feature on your calculator to determine that $n + 1 = 27$ so $n = 26$.

Chapter 16

1. $\lim_{n \to \infty} \left| \frac{(4x)^{n+2}}{n+2} \times \frac{n+1}{(4x)^{n+1}} \right| = |4x|$. If $|4x| < 1$, then $\frac{-1}{4} < x < \frac{1}{4}$. The series diverges when $x = \frac{1}{4}$ and converges for $x = \frac{-1}{4}$. The interval of convergence is $\frac{-1}{4} \le x < \frac{1}{4}$.

2. $f(0) = 1$, $f'(0) = 0$, $f''(0) = -1$, $f'''(0) = 0$. These numbers repeat for every fourth term. Each term in the expansion is of the form $\frac{f^{(n)}(0)x^n}{n!}$ so the series is $1 - \frac{x^2}{2} + \frac{x^4}{4!} - \frac{x^6}{6!} + \dots + \frac{(-1)^n x^{2n}}{(2n)!}$.

3. $\int 1 - \frac{x^2}{2} + \frac{x^4}{4!} - \frac{x^6}{6!} + \dots + \frac{(-1)^n x^{2n}}{(2n)!} \, dx = x - \frac{x^3}{3!} + \frac{x^5}{5!} - \frac{x^7}{7!} + \dots + \frac{(-1)^n x^{2n+1}}{(2n+1)!}$, which is the MacLaurin Series for $\sin(x)$, as expected.

4. $f(1) = 0$, $f'(1) = 1$, and $f''(1) = -1$, $f'''(1) = -2$. The Taylor Series for $f(x) = \ln(x)$ about $x = 1$

is $\sum_{n=1}^{\infty} \frac{f^{(n)}(1)(x-1)^n}{n!} = 1(x-1) - \frac{(x-1)^2}{2} + \frac{(x-1)^3}{3} - \frac{(x-1)^4}{4} + \dots + \frac{(-1)^{n+1}(x-1)^n}{n}$.

5. $\int_0^{1/2} 1 - x^4 + x^8 \, dx = x - \frac{x^5}{5} + \frac{x^9}{9} \Big|_0^{1/2} = 0.493967$. The difference $\left| \int_0^{1/2} f(x)\,dx - 0.493967 \right|$ is less than $\frac{\left(\frac{1}{2}\right)^{13}}{13} = 0.000009 < \frac{1}{10000}$.

Integration Practice Problems and Solutions

Here's your chance to go through a series of integration problems without knowing the chapter topic (which can hint at how to solve the problem). In this appendix, I provide a set of problems. Once you've worked through them, you can check your work against the solutions provided later in this appendix.

Good luck, and have fun with what you've learned!

Problems

1. $\int \frac{\left(\tan^{-1}(x)\right)^5}{1+x^2}\,dx$

2. $\int \frac{\sec(x)\tan(x)}{2\sec(x)-1}\,dx$

3. $\int \frac{x+1}{x^2+1}\,dx$

4. $\int \frac{1}{x\sqrt{1-\left(\ln(2x)\right)^2}}\,dx$

5. $\int x^3 e^{-2x}\,dx$

6. $\int \tan^3(x)\,dx$

7. $\int \left(\cos(x)+\sin(x)\right)^2 \cos(2x)\,dx$

8. $\int \sin^6(x)\cos^3(x)\,dx$

9. $\int_0^{\sqrt{2}} \frac{x}{4+x^4}\,dx$

10. $\int \frac{1}{\sqrt{x}+\sqrt{x^3}} dx$

11. $\int \tan^{-1}(x) dx$

12. $\int \frac{1}{\sqrt{x}+x} dx$

13. $\int x \sin^2(4x) dx$

14. $\int \frac{2x-1}{x^3-2x^2+x-2} dx$

Consider the graph of $f(x) = 2x^2$ and the y-axis on the closed interval $[1,4]$. Use this information to answer Problems 15 through 17.

15. Determine the length of the arc of the curve in this interval. Let R be the region bounded by $f(x)$ and the x-axis on the same interval. A solid is formed by rotating R around the x-axis.

16. Find the volume, V, of the solid.

17. Find the surface area, S, of the solid. Let R be the first quadrant region bounded by the graphs of $f(x) = \sin(x)$, $g(x) = \cos(x)$, and the y-axis. Use this information for Problems 18 through 20.

18. A solid with base R has cross sections perpendicular to the x-axis in the shape of squares. Find the volume of the solid.

19. Let S be the solid formed when R is rotated around the y-axis. Find the volume of S.

20. Let T be the solid when R is rotated around the x-axis. Find the volume of T.

Solutions

Let's see how you did.

1. $\int \frac{\left(\tan^{-1}(x)\right)^5}{1+x^2} dx = \frac{1}{6}\left(\tan^{-1}(x)\right)^6 + C$

   The derivative of $\tan^{-1}(x)$ is $\frac{1}{1+x^2}$, so let $u = \tan^{-1}(x)$ and $du = \frac{1}{1+x^2} dx$. Substitution gives $\int u^5 du = \frac{1}{6}u^6 + C$.

2. $\int \frac{\sec(x)\tan(x)}{2\sec(x)-1} dx = \frac{1}{2}\ln\left|2\sec(x)-1\right| + C$

   The derivative of $2\sec(x) - 1$ is $\sec(x)\tan(x)$, so let $u = 2\sec(x) - 1$ and $du = \sec(x)\tan(x)$ dx. Substitution gives $\frac{1}{2}\int \frac{1}{u} du = \frac{1}{2}\ln|u| + C$.

3. $\int \frac{x+1}{x^2+1} dx = \frac{1}{2} \ln|x^2+1| + \tan^{-1}(x) + C$

Rewrite $\int \frac{x+1}{x^2+1} dx$ as $\int \frac{x}{x^2+1} + \frac{1}{x^2+1} dx$. The derivative of x^2+1 is $2x$, so the numerator of the first fraction is half the derivative of the denominator. This means $\int \frac{x}{x^2+1} dx = \frac{1}{2} \ln|x^2+1|$ and that the second part of the integrand is the formula for the antiderivative of $\tan^{-1}(x)$.

4. $\int \frac{1}{x\sqrt{1-(\ln(2x))^2}} dx = \sin^{-1}(\ln(2x)) + C$

The key to this problem is the radical in the denominator contains the difference of two squares. If you let $u = \ln(2x)$, then $du = \frac{2}{2x} dx = \frac{1}{x} dx$. Substitution gives $\int \frac{1}{\sqrt{1-u^2}} du = \sin^{-1}(u) + C$.

5. $\int x^3 e^{-2x} dx = \frac{-1}{2} x^3 e^{-2x} - \frac{3}{4} x^2 e^{-2x} - \frac{3}{4} xe^{-2x} - \frac{3}{8} e^{-2x} + C$

The polynomial is not the derivative of the exponent in the exponential function. Let's use the Tabular Method to solve this problem.

u	dv	+1
x^3	e^{-2x}	1
$3x^2$	$\frac{-1}{2} e^{-2x}$	-1
$6x$	$\frac{1}{4} e^{-2x}$	1
6	$\frac{-1}{8} e^{-2x}$	-1
0	$\frac{1}{16} e^{-2x}$	1
		-1

Figure B.1

$\int x^3 e^{-2x} dx = (x^3)(\frac{-1}{2} e^{-2x})(1) + (3x^2)(\frac{1}{4} e^{-2x})(-1) + (6x)(\frac{-1}{8} e^{-2x})(1) + (6)(\frac{1}{16} e^{-2x})(-1)$

$\int x^3 e^{-2x} dx = \frac{-1}{2} x^3 e^{-2x} - \frac{3}{4} x^2 e^{-2x} - \frac{3}{4} xe^{-2x} - \frac{3}{8} e^{-2x} + C$

6. $\int \tan^3(x)\,dx = \frac{1}{2}\tan^2(x) - \ln\left|\sec(x)\right| + C$

Rewrite $\tan^3(x)$ as $\tan(x)\tan^2(x)$ and then use the Pythagorean identity $\tan 2(x) = \sec^2(x) - 1$. $\int \tan^3(x)\,dx$ becomes $\int \tan(x)\sec^2(x) - \tan(x)\,dx$. With $\sec^2(x)$ being the derivative of $\tan(x)$, the antiderivative of $\tan(x)\sec^2(x)$ is $\frac{1}{2}\tan^2(x)$. Earlier, we saw that the antiderivative of $\tan(x)$ is $\ln|\sec(x)|$.

7. $\int \left(\cos(x) + \sin(x)\right)^2 \cos(2x)\,dx = \frac{1}{2}\sin\left(2x\right) + \frac{1}{4}\sin^2(2x) + C$

$(\cos(x) + \sin(x))^2 = \cos^2(x) + 2\cos(x)\sin(x) + \sin^2(x)$. Since $\cos^2(x) + \sin^2(x) = 1$ and $2\cos(x)\sin(x) = \sin(2x)$, the integrand is now $\cos(2x)(1 + \sin(2x))$, which expands to $\cos(2x) + \cos(2x)\sin(2x)$. The antiderivative for $\cos(2x)$ is $\frac{1}{2}\sin(2x)$. The second piece of the integrand, $\cos(2x)\sin(2x)$, offers you two options, each is correct:

- If $u = \cos(2x)$, then $du = -2\sin(2x)dx$ and the problem becomes $\frac{-1}{2}\int u\,du = \frac{-1}{4}u^2$, which translates back to $\frac{-1}{4}\cos^2(2x)$.

- If $u = \sin(2x)$, then $du = 2\cos(2x)dx$. The resulting integral would be $\frac{1}{4}\sin^2(2x)$.

(This is another example of how the constant of integration absorbs any extra constants that are floating around. $\frac{-1}{4}\cos^2(2x) = \frac{-1}{4}\left(1 - \sin^2(2x)\right) = \frac{-1}{4} + \frac{1}{4}\sin^2(2x)$. The extra $\frac{-1}{4}$ is absorbed by C.)

8. $\int \sin^6(x)\cos^3(x)\,dx = \frac{1}{7}\sin^7(x) - \frac{1}{9}\sin^9(x) + C$

Use $\cos^3(x) = \cos^2(x)\cos(x) = \cos(x)(1 - \sin^2(x))$. The integrand is now $\sin^6(x)\cos(x) - \sin^8(x)\cos(x)$.

9. $\int_0^{\sqrt{2}} \frac{x}{4+x^4}\,dx = \frac{\pi}{16}$

Rewrite x^4 as $(x^2)^2$. The integrand is the sum of two squares leading us to believe this is some type of $\tan^{-1}(x)$ problem. Factor 4 from the denominator, $\frac{1}{4}\int \frac{x}{1+\left(\frac{x^2}{2}\right)^2}\,dx$.

Let $u = \frac{x^2}{2}$, so that $du = xdx$. Substitution gives $\frac{1}{4}\int \frac{1}{1+u^2}\,du = \frac{1}{4}\tan^{-1}(u)$. Rewrite the result in terms of the original variable to get $\int_0^{\sqrt{2}} \frac{x}{4+x^4}\,dx = \frac{1}{4}\tan^{-1}\left(\frac{x^2}{2}\right)\Big|_0^{\sqrt{2}} = \frac{1}{4}\left(\tan^{-1}(1) - \tan^{-1}(0)\right) = \frac{1}{4}\left(\frac{\pi}{4} - 0\right) = \frac{\pi}{16}$.

10. $\int \frac{1}{\sqrt{x}+\sqrt{x^3}}dx = 2\tan^{-1}\left(\sqrt{x}\right)+C$

The denominator has a common factor of $\sqrt{x}$. $\sqrt{x}+\sqrt{x^3} = \sqrt{x}\left(1+\sqrt{x^2}\right) = \sqrt{x}\left(1+\left(\sqrt{x}\right)^2\right)$.
What you need to see here is that the denominator contains the sum of squares. Let

$u = \sqrt{x} = x^{1/2}$ then $du = \frac{1}{2}x^{-1/2}dx$, so $2du = x^{-1/2}dx$. Substitution gives

$2\int \frac{1}{1+u^2}dx = 2\tan^{-1}(u) + C$. Working back to the original variable, $\int \frac{1}{\sqrt{x}+\sqrt{x^3}}dx =$
$2\tan^{-1}\left(\sqrt{x}\right)+C$.

11. $\int \tan^{-1}(x)dx = x\tan^{-1}(x) - \frac{1}{2}\ln\left|x^2+1\right| + C$

This one is a little tricky. The answer is not $\frac{1}{1+x^2}$. ($\frac{1}{1+x^2}$ is the derivative of $\tan^{-1}(x)$ not
the antiderivative.) Use integration by parts with $u = \tan^{-1}(x)$ and $dv = dx$. You get $du = \frac{1}{1+x^2}dx$ and $v = x$.

$\int \tan^{-1}(x)dx = x\tan^{-1}(x) - \int \frac{x}{1+x^2}dx$. As we saw in Problem 3, $\int \frac{x}{x^2+1}dx = \frac{1}{2}\ln\left|x^2+1\right|$
Consequently, $\int \tan^{-1}(x)dx = x\tan^{-1}(x) - \frac{1}{2}\ln\left|x^2+1\right| + C$.

12. $\int \frac{1}{\sqrt{x}+x}dx = 2\ln\left|1+\sqrt{x}\right|+C$

This looks like Problem 10, which might get you to thinking about factoring $\sqrt{x}$ from
the denominator. $\sqrt{x}+x = \sqrt{x}\left(1+\sqrt{x}\right)$. Let $u = 1+\sqrt{x}$ so that $du = \frac{1}{2}x^{-1/2}dx$ and
$2\,du = x^{-1/2}dx$. Substitution gives $2\int \frac{1}{u}du = 2\ln|u|$. Going back to the original variable,
$\int \frac{1}{\sqrt{x}+x}dx = 2\ln\left|1+\sqrt{x}\right|+C$.

13. $\int x\sin^2(4x)dx = \frac{1}{4}x^2 - \frac{1}{16}x\sin(8x) + \frac{1}{128}\cos(8x)+C$

The product of the polynomial and trigonometric function should lead us to use
integration by parts. Let $u = x$ and $dv = \sin^2(4x)dx = \frac{1-\cos(8x)}{2}dx$, so that $du = dx$ and $v = \frac{1}{2}x - \frac{1}{16}\sin(8x)$.

$\int x\sin^2(4x)dx = x\left(\frac{1}{2}x - \frac{1}{16}\sin(8x)\right) - \int \frac{1}{2}x - \frac{1}{16}\sin(8x)dx =$
$\frac{1}{2}x^2 - \frac{1}{16}x\sin(8x) - \frac{1}{4}x^2 + \frac{1}{128}\cos(8x) + C = \frac{1}{4}x^2 - \frac{1}{16}x\sin(8x) + \frac{1}{128}\cos(8x)+C$

14. $\int \frac{2x-1}{x^3-2x^2+x-2}dx = \frac{3}{5}\ln|x-2| - \frac{3}{10}\ln|x^2+1| + \frac{4}{5}\tan^{-1}(x) + C$

The denominator of the integrand factors to $(x-2)(x^2+1)$ leads us to use partial fractions as the technique to solve this problem.

$$\frac{2x-1}{x^3-2x^2+x-2} = \frac{A}{x-2} + \frac{Bx+C}{x^2+1}$$

Multiply through by the common denominator to get:

$$2x-1 = A(x^2+1) + (Bx+C)(x-2)$$

Set $x = 2$: $3 = A(5)$ so $A = \frac{3}{5}$.

Set $x = 0$: $-1 = \frac{3}{5} + C(-2)$ so that $\frac{-8}{5} = -2C$ and $C = \frac{4}{5}$.

Let $x = 1$: $1 = \frac{3}{5}$ $(2) + (B + \frac{4}{5})(-1)$, which gives $1 = \frac{6}{5} - B - \frac{4}{5} = \frac{2}{5} - B$, so $B = \frac{-3}{5}$.

$\int \frac{2x-1}{x^3-2x^2+x-2}dx = \frac{3}{5}\int \frac{1}{x-2}dx + \frac{1}{5}\int \frac{-3x+4}{x^2+1}dx = \frac{3}{5}\int \frac{1}{x-2}dx - \frac{3}{5}\int \frac{x}{x^2+1}dx + \frac{4}{5}\int \frac{1}{x^2+1}dx =$

$\frac{3}{5}\ln|x-2| - \frac{3}{10}\ln|x^2+1| + \frac{4}{5}\tan^{-1}(x) + C$

Consider the graph of $f(x) = 2x^2$ and the y-axis on the closed interval $[1,4]$. Use this information to answer Problems 15 through 17.

15. Determine the length of the arc of the curve in this interval.

The derivative of $f(x) = 4x$, so the length of the arc is $L = \int_1^4 \sqrt{1+16x^2}\,dx$. The sum of the squares within the radical is an indication that trigonometric substitution is a viable option.

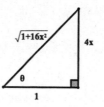

Figure B.2

Let $\sec(\theta) = \sqrt{1+16x^2}$ and $\tan(\theta) = 4x$ so that $dx = \frac{1}{4}\sec^2(\theta)d\theta$.

$\int \sqrt{1+16x^2}\,dx$ becomes $\frac{1}{4}\int \sec^3(\theta)d\theta = \frac{1}{4}\left(\frac{1}{2}\sec(\theta)\tan(\theta) + \frac{1}{2}\ln|\sec(\theta) + \tan(\theta)|\right)$.

Returning to the original variable, $\int_1^4 \sqrt{1+16x^2}\, dx =$

$$\frac{1}{8}\left(4x\sqrt{1+16x^2}+\ln\left(4x+\sqrt{1+16x^2}\right)\right)\Big|_1^4 =$$

$\frac{1}{8}\left[\left(16\sqrt{257}+\ln\left(16+\sqrt{257}\right)\right)-\left(4\sqrt{17}+\ln\left(4+\sqrt{17}\right)\right)\right]$. In more manageable (meaningful?) numbers, the arc length is 30.1724.

Let R be the region bounded by $f(x)$ and the x-axis on the same interval. A solid is formed by rotating R about the x-axis.

16. Find the volume, V, of the solid.

A cross section of the solid formed is a disk and the radius of this disk will be $f(x)$. Therefore, the volume of the solid formed is

$$\pi\int_1^4\left(2x^2\right)^2 dx = \pi\int_1^4 4x^4\, dx = \pi\left(\frac{4}{5}x^5\right)\Big|_1^4 = \pi\left(\frac{4096}{5}-\frac{4}{5}\right) = \frac{4092\pi}{5}.$$

17. Find the surface area, S, of the solid.

The surface area of the solid is $2\pi\int_1^4 2x^2\sqrt{1+16x^2}\, dx$. Using Figure B.2, let $\sec(\theta) = \sqrt{1+16x^2}$ and $\frac{1}{4}\tan(\theta) = x$ so that $dx = \frac{1}{4}\sec^2\left(\theta\right)d\theta$.

Converting the integral to trigonometric functions, $2\pi\int_1^4 2x^2\sqrt{1+16x^2}\, dx =$

$\frac{\pi}{16}\int \tan^2\left(\theta\right)\sec^3\left(\theta\right)d\theta = \frac{\pi}{16}\int\left(\sec^2\left(\theta\right)-1\right)\sec^3\left(\theta\right)d\theta = \frac{\pi}{16}\int\left(\sec^5\left(\theta\right)-\sec^3\left(\theta\right)\right)d\theta =$

$\frac{\pi}{64}\left[\sec^3\left(\theta\right)\tan\left(\theta\right)-\frac{1}{2}\sec\left(\theta\right)\tan\left(\theta\right)-\frac{1}{2}\ln\left|\sec\left(\theta\right)+\tan\left(\theta\right)\right|\right]$. Return to the original variable, $2\pi\int_1^4 2x^2\sqrt{1+16x^2}\, dx =$

$\frac{\pi}{64}\left[4x\left(\sqrt{1+16x^2}\right)^3-2x\sqrt{1+16x^2}-\frac{1}{2}\ln\left|\sqrt{1+16x^2}+4x\right|\right]\Big|_1^4 = 3216.2.$

Let R be the first quadrant region bounded by the graphs of $f(x) = \sin(x)$, $g(x) = \cos(x)$, and the y-axis. Use this information for Problems 18 through 20.

The region R is:

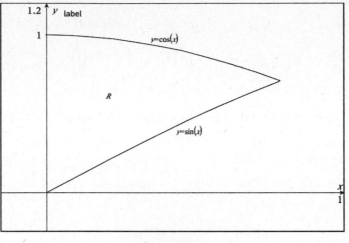

Figure B.3

18. A solid with base R has cross sections perpendicular to the x-axis in the shape of squares. Find the volume of the solid.

The length of a side of the square is $\cos(x) - \sin(x)$. Therefore, the volume of the solid formed is $\int_0^{\pi/4} \left(\cos(x) - \sin(x)\right)^2 dx = \int_0^{\pi/4} \cos^2(x) - 2\sin(x)\cos(x) + \sin^2(x)\,dx =$

$2\pi\left[\left(\left(\tfrac{1}{4}\right)\left(\tfrac{\sqrt{2}}{2}\right)+\left(\tfrac{1}{4}\right)\left(\tfrac{\sqrt{2}}{2}\right)+\tfrac{\sqrt{2}}{2}-\tfrac{\sqrt{2}}{2}\right)-(1)\right] = x + \tfrac{1}{2}\cos(2x)\Big|_0^{\pi/4} = \left(\tfrac{\pi}{4} + \tfrac{1}{2}\cos\left(\tfrac{\pi}{2}\right)\right) - \left(0 + \tfrac{1}{2}\cos(0)\right) = \tfrac{\pi}{4} - \tfrac{1}{2}.$

19. Let S be the solid formed when R is rotated about the y-axis. Find the volume of S.

Because we are revolving R around a vertical line, let's use cylindrical shells. The radius of each cylinder is x, and the height of the cylinder is $\cos(x) - \sin(x)$. Therefore, the volume of S is $2\pi \int_0^{\pi/4} x(\cos(x) - \sin(x))\,dx = 2\pi \int_0^{\pi/4} x\cos(x) - x\sin(x)\,dx$.

Each of these terms can be integrated using integration by parts.

For $x\cos(x)$, let $u = x$ and $dv = \cos(x)\,dx$, so that $du = dx$ and $v = \sin(x)$.

$$\int x\cos(x)\,dx = x\sin(x) - \int \sin(x)\,dx = x\sin(x) + \cos(x)$$

For $x\sin(x)$, let $u = x$ and $dv = \sin(x)\,dx$, so that $du = dx$ and $v = -\cos(x)$.

$$\int x\sin(x)\,dx = -x\cos(x) + \int \cos(x)\,dx = -x\cos(x) + \sin(x)$$

Therefore, $2\pi \int_0^{\pi/4} x\cos(x) - x\sin(x)\, dx = 2\pi \left(x\sin(x) + x\cos(x) + \cos(x) - \sin(x) \right)\Big|_0^{\pi/4} =$

$2\pi\left[\left(\frac{\pi}{4}\sin\left(\frac{\pi}{4}\right) + \frac{\pi}{4}\cos\left(\frac{\pi}{4}\right) - \sin\left(\frac{\pi}{4}\right) + \cos\left(\frac{\pi}{4}\right)\right) - \left(0\sin(0) + 0\cos(0) + \cos(0) - \sin(0)\right)\right] =$

$2\pi\left[\left(\left(\frac{\pi}{4}\right)\left(\frac{\sqrt{2}}{2}\right) + \left(\frac{\pi}{4}\right)\left(\frac{\sqrt{2}}{2}\right) + \frac{\sqrt{2}}{2} - \frac{\sqrt{2}}{2}\right) - \left(1\right)\right] = 2\pi\left(\frac{\pi\sqrt{2}}{4} - 1\right)$.

20. Let T be the solid when R is rotated around the x-axis. Find the surface area of T.

    A cross section of T shows a disk with larger radius $\cos(x)$ and smaller radius $\sin(x)$.
    The volume of T is $\pi\int_0^{\pi/4}\cos^2(x) - \sin^2(x)\, dx$. Use the original trigonometric identity
    for $\cos(2x) = \cos^2(x) - \sin^2(x)$ to get $\pi\int_0^{\pi/4}\cos^2(x) - \sin^2(x)\, dx = \pi\int_0^{\pi/4}\cos(2x)\, dx =$
    $\frac{\pi}{2}\sin(2x)\Big|_0^{\pi/4} = \frac{\pi}{2}\left(\sin\left(\frac{\pi}{2}\right) - \sin(0)\right) = \frac{\pi}{2}$.

Glossary

absolute convergence   Describes when the series $\sum_{k=1}^{\infty} |a_k|$ converges. If $\sum_{k=1}^{\infty} |a_k|$ converges, then so does $\sum_{k=1}^{\infty} a_k$.

acceleration   The rate of change of the velocity.

alternating series   A series whose consecutive terms alternate between positive and negative values.

Alternating Series Test   Let $a_n = \left(-1\right)^n b_n$ with $b_n > 0$ for all n. If b_n is decreasing for all n, and $\lim_{n \to \infty} b_n = 0$ then $\sum_{k=1}^{\infty} a_k$ is convergent.

antiderivative   The expression from which the derivative was found.

average value of a function   The average value of a function f(x) on the interval [a,b] is $\frac{1}{b-a} \int_a^b f(x)\,dx$.

Chain Rule   The rule used for differentiating functions that are composed of other functions. If f(x) = g(k(x)), then f'(x) = g'(k(x))k'(x).

Comparison Test   If two positive series, $\sum_{k=1}^{\infty} a_k$ and $\sum_{n=1}^{\infty} b_n$, with the property that $a_k > b_k$ for $k \geq m$, then we can conclude that if the series $\sum_{k=1}^{\infty} a_k$ converges, then so does the series $\sum_{n=1}^{\infty} b_n$. Also, if the series $\sum_{n=1}^{\infty} b_n$ diverges, then so does the series $\sum_{k=1}^{\infty} a_k$.

Comparison Test for Improper Integrals   If f(x) and g(x) are continuous functions for $x > a$ with f(x) > g(x) > 0, then (1) If $\int_a^{\infty} f(x)\,dx$ converges, then so does $\int_a^{\infty} g(x)\,dx$. (2) If $\int_a^{\infty} g(x)\,dx$ diverges, then so does $\int_a^{\infty} f(x)\,dx$.

conditionally convergent   A series that is convergent but not absolutely convergent.

constant of integration   The constant added to the end of an indefinite integral to indicate there is a family of functions that satisfy the given integral.

continuous at a point   A function is continuous at a point $x = c$ if and only if $\lim_{x \to c} f(x) = f(c)$.

continuous function   A function that is continuous at all points in its domain.

converge   A sequence or series that is bounded by some finite value.

decreasing sequence   A sequence in which $a_{n+1} \leq a_n$ for all n.

definite integral   An integral that contains a lower and upper bound and whose answer is a real number.

differential equation   An equation that contains a derivative.

disk method   The method used to calculate the volume of a solid of rotation when the interior of the solid contains no holes.

displacement   The change in position of an object from the start to an end of its travel. This is not necessarily the same as the distance traveled.

Divergence Test   The infinite series $\displaystyle\sum_{k=1}^{\infty} a_k$ is divergent if $\lim_{n \to \infty} a_n \neq 0$.

divergent   A sequence or series that does not have a limiting value.

dot product   Given two vectors $a = (a_1, a_2)$ and $b = (b_1, b_2)$, the dot product is $\mathbf{a} \cdot \mathbf{b} = a_1 b_1 + a_2 b_2$.

Euler's Method   A technique used to approximate a value of a function using a known value of the function and the derivative of the function.

exponential growth   A process in which the rate of growth of a function is proportional to the current value of the function.

First Order Linear Differential Equation   A differential equation of the form $\frac{dy}{dt} + P(t)y = Q(t)$.

Fundamental Theorem of Calculus   Given the function F(x) whose derivative is f(x), the Fundamental Theorem of Calculus says that $\int_a^b f(x)\,dx = F(b) - F(a)$.

geometric sequence   A sequence in which the ratio of consecutive terms is a constant.

geometric series   The sum of the terms of a geometric sequence.

implicit differentiation   When equations cannot be written in the form $y = f(x)$, the notation $\frac{dy}{dx}$ indicates that it is implied that y is a function of x so that whenever a term in y is differentiated, the Chain Rule needs to be applied by attaching the term $\frac{dy}{dx}$ to it.

improper integral   An integral that has at least one of its bounds going to infinity or an integral that contains an infinite discontinuity within the bounds of integration.

increasing sequence   A sequence in which $a_{n+1} \geq a_n$ for all n.

indefinite integral   An integral that yields a family of functions each of whose derivative is the integrand of the integral.

infinite discontinuity   Discontinuity caused by a vertical asymptote.

integrand   The expression contained within the integral whose antiderivative is being sought. $f(x)$ is the integrand in the $\int f(x)\,dx$.

integration by parts   A consequence of the product rule for differentiation: $d(uv) = udv + vdu$ becomes $udv = d(uv) - vdu$ so that $\int u\,dv = uv - \int v\,du$.

Integration Test   If $f(x)$ is a continuous, positive, decreasing function on $[1,\infty]$ with $f(n) = a_n$ for all positive integers n, then the series $\sum a_n$ converges if and only if $\int_1^\infty f(x)\,dx$ converges.

interval of convergence   The interval on which a power series converges and goes from $c - r$ to $c + r$ where c is the point about which the power series was constructed and r is the radius of convergence.

Lagrange Error Estimate (or Lagrange Remainder)   Given a Taylor Series $T(x)$ for $f(x)$, the Lagrange Remainder, $R(x)$, for the difference $f(x) - T(x)$ is $R(x) = \frac{f^{(n+1)}(x^*)}{(n+1)!}\left(x - x_0\right)^{n+1}$ where $f^{(n+1)}(x^*)$ is the nth derivative evaluated at the point x^*.

L'Hopital's Rule   Given $\lim\limits_{x \to c} \frac{f(x)}{g(x)}$, with both $f(x)$ and $g(x)$ differentiable at $x = c$. If $\lim\limits_{x \to c} \frac{f(x)}{g(x)} = \frac{0}{0}$ or $\pm\frac{\infty}{\infty}$, then $\lim\limits_{x \to c} \frac{f(x)}{g(x)} = \lim\limits_{x \to c} \frac{f(x)}{g'(x)}$.

Limit Comparison Test   Given two positive series $\sum\limits_{k=1}^{\infty} a_k$ and $\sum\limits_{n=1}^{\infty} b_n$, let $\lim\limits_{n \to \infty} \frac{a_n}{b_n} = L$. If L is positive and finite, then both $\sum\limits_{k=1}^{\infty} a_k$ and $\sum\limits_{n=1}^{\infty} b_n$ converge or both diverge.

logistic growth   A phenomenon which looks like exponential growth but eventually slows and reaches a plateau.

MacLaurin Series   A series of the form $\sum\limits_{n=0}^{\infty} \frac{f^{(n)}(0)x^n}{n!}$ that approximates a function $f(x)$ about the point $x = 0$.

Mean Value Theorem   If a function $f(x)$ is defined and continuous on $[a,b]$ and differentiable of (a,b), there is a value of c in the interval (a,b) so that the instantaneous rate of change at c is equal to the average rate of change over $[a,b]$. That is, there exists a value of c in (a,b) so that $f'(c) = \frac{f(b) - f(a)}{b - a}$.

p-series   A series of the form $\sum\limits_{n=1}^{\infty} \frac{1}{n^p}$. The series converges when $p > 1$ and diverges when $p \leq 1$.

parametric equations   The coordinates x and y are each written in terms of a third parameter, usually t.

partial fraction decomposition   A technique used to separate a fraction into the sum and difference of smaller fractions.

partition   A partition is a subset of the interval $[a,b]$ on which an integral is being computed.

piece-wise (split domain) function   A function that is defined by different rules for specific subsets of the domain is called a piece-wise function.

polar coordinates   A coordinate system in which the location of a point is determined by the radius of a circle drawn from a fixed point and the measure of an angle drawn from a fixed ray.

positive series   A series that only contains positive terms.

power series   A series centered at $x = c$ that has the form $\sum_{n=0}^{\infty} a_n (x - c)^n$.

Product Rule   If $f(x) = g(x) \times k(x)$, then $f(x) = g(x)k'(x) + g'(x)k(x)$.

Quotient Rule   If $f(x) = \frac{g(x)}{k(x)}$, then $f(x) = \frac{g'(x)k(x) - g(x)k'(x)}{(k(x))^2}$.

radius of convergence   The distance from a central value about which a power series will converge. *See also* interval of convergence.

Ratio Test   Given a series $\sum_{k=1}^{\infty} a_k$, let $\lim_{n \to \infty} \left| \frac{a_{n+1}}{a_n} \right| = r$. If $r < 1$, then $\sum_{k=1}^{\infty} a_k$ converges. If $r > 1$, then $\sum_{k=1}^{\infty} a_k$ diverges. If $r = 1$, then no conclusion about $\sum_{k=1}^{\infty} a_k$ can be drawn from this test.

Riemann Sum   A Riemann Sum takes the interval $[a,b]$, divides it into a number of partitions, and computes the area under the curve for each partition for the purpose of estimating the integral $\int_a^b f(x)\, dx$.

Root Test   Given a series $\sum_{k=1}^{\infty} a_k$, let $\lim_{n \to \infty} \sqrt[n]{|a_n|} = r$. If $r < 1$, then $\sum_{k=1}^{\infty} a_k$ is absolutely convergent. If $r > 1$, then $\sum_{k=1}^{\infty} a_k$ divergent. If $r = 1$, then no conclusion about $\sum_{k=1}^{\infty} a_k$ can be drawn from this test.

scalar   A quantity that has magnitude only.

Second Fundamental Theorem of Calculus   If $Q(x) = \int_{g(x)}^{k(x)} f(t)\, dt$, then $Q'(x) = f(k(x))k'(x) - f(g(x))g'(x)$.

separation of variables   A technique used to solve simple differentiable equations. All terms of a given variable are moved to one side of the equation.

sequence   A listing of numbers generated by a mathematical rule.

series   The sum of the terms in a sequence.

shell method   A procedure used to compute the volume of a solid of rotation by accumulating the surface area of a shell and multiplying it by the thickness.

Simpson's Rule   Simpson's Rule uses the area under a parabolic arc determined by three data points to approximate the area under a curve.

slope field   A visual representation of a differential equation. It is traditional to use lattice points (points whose coordinates are integers), compute the slope of the tangent line to the function using the differential equation, and draw a small line segment with that slope at that point.

Squeeze Theorem   If $a_n \leq b_n \leq c_n$ for $n \geq k$, where k is some constant, with $\lim\limits_{n \to \infty} a_n = L$ and $\lim\limits_{n \to \infty} c_n = L$, then $\lim\limits_{n \to \infty} b_n = L$.

strictly decreasing sequence   A sequence in which $a_{n+1} < a_n$ for all n.

strictly increasing sequence   A sequence in which $a_{n+1} > a_n$ for all n.

Taylor Series   A series of the form $\sum\limits_{n=0}^{\infty} \frac{f^{(n)}(c)(x-c)^n}{n!}$ that approximates a function $f(x)$ about a point $x = c$.

transcendental function   A function that cannot be written as a polynomial or ratio of polynomials.

trapezoidal rule   The trapezoidal approximation of a Riemann Sum has the height of each trapezoid as the width of the interval, while the bases of the trapezoid are the functional values at each endpoint.

***u*-substitution**   A technique used when both a function and its derivative appear in the integrand.

vector   A quantity that has both magnitude and direction.

washer method   The method used to calculate the volume of a solid of rotation when the interior of the solid contains holes.

Index

E

F

M

MacLaurin Series
  interval of convergence, 256-258
  Lagrange Error Estimate, 260-262
  sample problems, 252-258
Mean Value Theorem
  overview, 29-30
  true area calculations, 76
midpoint method (Riemann Sums), 70
monotonic sequences
  overview, 233
  sample problems, 234
multiplication, vectors
  cross products, 199
  dot products, 199-200

N–O

natural numbers, 257
Newton, Isaac, 24
nonrepeating linear factors, partial fractions, 152-155
numbers
  complex, 257
  imaginary, 257
  irrational, 257
  natural, 257
  rational, 257
  real, 257
  whole, 257

odd powers
  sine and tangent integral form, 139-141
  tangent and secant integral form, 143-146

P

parabolas
  defined parametrically, 9
  limits, 20
parabolic arcs, Simpson's Rule, 87
parallelogram method, adding vectors, 197
parametric equations, 171
  overview, 8-10, 172
  parametric curves
    arc lengths, 178-179
    first derivatives, 172-176
    second derivatives, 177
partial fractions
  integration
    nonrepeating linear factors, 152-153
    quadratic factors, 155-157
    repeated linear factors, 154-155
  overview, 15-17
particular solution, differential equations, 206
phase shift, 5
piecewise function, 21
planes
  Cartesian Coordinate plane, 10
  polar coordinates
    cardioid graph, 12
    graphing, 12-14
    limacon graph, 12
    overview, 10-14
    rose graph, 12
point of inflection, 31
polar coordinates
  graphs, 12-14
    cardioid, 12
    limacon, 12
    rose, 12

Q-R

U

V

W–X–Y–Z